FEDERAL DISASTER PROGRAMS AND HURRICANE KATRINA

FEDERAL DISASTER PROGRAMS AND HURRICANE KATRINA

DOUGLAS D. SYZERHANS
EDITOR

Nova Science Publishers, Inc.
New York

For permission to use material from this book please contact us:
Telephone 631-231-7269; Fax 631-231-8175
Web Site: http://www.novapublishers.com

LIBRARY OF CONGRESS CATALOGING-IN-PUBLICATION DATA

Federal disaster programs and Hurricane Katrina / Douglas D. Syzerhans (editor).
p. cm.
Includes bibliographical references and index.
ISBN 1-59454-889-7
1. Hurricane Katrina, 2005. 2. Disaster relief--Government policy --United States. 3. Emergency management--Government policy--United States. I. Syzerhans, Douglas D.
HV636 2005 .U6 F43 2006
363.34'922560973--dc22 2005033113

Published by Nova Science Publishers, Inc. ✦New York

CONTENTS

PREFACE

Federal disaster programs kind of sit there and gather moss after all the expenses of running the government agencies responsible for rendering assistance to the disaster victims are spent on staff salaries, computers, travel and all the accompaniments of perceived power. Hopefully, there will be no disasters. Otherwise, let them be small disasters which might not interrupt lunch. In the case of Hurricane Katrina, the federal disaster programs were themselves disasters which were limp responses to the thousands of dead, hundreds of thousands homeless and entire sectors of America destroyed. This new book tries to examine the initial disaster programs, recovery disaster programs designed to cover-up for the initial flaws and the programs planned to prevent more disastrous disaster programs.

In: Federal Disaster Programs and Hurricane Katrina
Editor: Douglas D. Syzerhans, pp. 1-6
ISBN 1-59454-889-7

Chapter 1

HURRICANE KATRINA: ACTIVITIES OF THE SOCIAL SECURITY ADMINISTRATION

Scott Szymendera

SUMMARY

Shortly after Hurricane Katrina hit Alabama, Louisiana, and Mississippi, the Social Security Administration (SSA) implemented procedures to assist affected beneficiaries. These procedures are designed to ensure an uninterrupted stream of retirement, disability and survivors benefits as well as Supplemental Security Income (SSI) benefits during the immediate aftermath of this crisis. This report will outline these procedures and also discuss SSA procedures after September 11, 2001, to facilitate new disability, SSI, survivors, and death benefit applications. This report provides links to additional information from SSA and other federal agencies. Because of the dynamic nature of the situation, staff are encouraged to contact the Congressional Research Service (CRS) or consult the links provided at the end of this report for updated information.

PROCEDURES FOR BENEFICIARIES

Over 1.2 million Social Security retirement and disability and survivors beneficiaries are in the counties affected by Hurricane Katrina.[1] In addition, this same area has just under 400,000 Supplemental Security Income (SSI) beneficiaries[2] as shown in Table 1.[3] According to SSA, ensuring that these beneficiaries are able to access their monthly benefits, whether via paper checks or direct deposit, is a key feature of the agency's response to this crisis.[4]

BENEFICIARIES WHO RECEIVE PAPER CHECKS

According to SSA, physical benefit checks have been and will be mailed on schedule to beneficiaries. However, because of interruptions in the mail delivery system and the relocation of residents from affected areas, many beneficiaries may not be able to receive their checks in a timely manner.[5] SSA, working with the United States Postal Service, has established the following procedures for beneficiaries unable to access paper checks through the mail:

- Beneficiaries have been able to pick up their original, re-routed, checks at designated Temporary Mail Delivery stations established by the United States Postal Service. All Temporary Mail Delivery stations will be closed as of September 16, 2005.[6]
- Beneficiaries who do not pick up their original checks at Temporary Mail Delivery stations may receive immediate payments at any open Social Security office in the country. A list of open offices in Alabama, Louisiana, and Mississippi is provided in Table 2;
- Checks that are not picked up at Temporary Mail Delivery stations will be forwarded if a change of address card has been filed with the United States Postal Service. The United States Postal Service is waiving the usual 30 day limit on address changes and allowing displaced persons to have mail forwarded to temporary shelters.

The Associated Press reports that, as of September 8, 2005, approximately 15,000 Social Security checks have been delivered at Temporary Mail Delivery stations.

Table 1. Social Security Beneficiaries in Affected Areas of Alabama, Louisiana, and Mississippi

State	Disability, Retirement, and Survivors	SSI	Total
Alabama (5 affected counties)	121,650	19,925	141,575
Louisiana (entire state)	740,378	169,547	909,925
Mississippi (51 affected counties)	386,675	202,907	589,582
Total	1,248,703	392,379	1,641,082

Source: Calculations of the Congressional Research Service (CRS) using December 2004 data from the Social Security Administration (SSA).

Note: Affected counties are those included in the Presidential Disaster Declaration as reported by the Federal Emergency Management Agency (FEMA).

Beneficiaries Who Receive Direct Deposit

Approximately 76% of Social Security retirement, disability and survivors beneficiaries in Alabama and Mississippi and 72% of Social Security retirement, disability and survivors beneficiaries in Louisiana have their benefits direct deposited into their checking accounts.[7] According to SSA, all direct deposits have been and will be made on schedule. However, due to bank closures and the relocation of persons from affected areas, beneficiaries may not be able to access their accounts and their deposited benefits. SSA has established the following procedure to provide benefits to those unable to access benefits deposited into their accounts:

- Beneficiaries unable to access benefits deposited into their accounts may receive immediate payments at any open Social Security office. A list of open offices in Alabama, Louisiana, and Mississippi is provided in Table 2. However, beneficiaries from these states may receive immediate payments at any Social Security office in the country.

Table 2. Status of SSA Field Offices in Affected States (as of September 8, 2005)

Alabama		**Louisiana**		**Mississippi**	
Office	**Status**	**Office**	**Status**	**Office**	**Status**
Albertville	Open	Alexandria	Open	*Biloxi*	*Closed*
Alexander City	Open	Bastrop	Open	Brookhaven	Open
Andalusia	Open	Baton Rouge North	Open	Clarksdale	Open
Anniston	Open	Baton Rouge Downtown	Open	Cleveland	Open
Bessemer	Open	*Bogalusa*	*Closed*	Columbus	Open
Birmingham (12th Avenue)	Open	*Covington*	*Closed*	Corinth	Open
Birmingham(Gadsen Highway)	Open	DeRidder	Open	Greenville	Open
Cullman	Open	Hammond	Open	Greenwood	Open
Decatur	Open	Houma	Open	Grenada	Open
Dothan	Open	*Kenner*	*Closed*	*Gulfport*	*Closed*
Fairhope	Open	Lafayette	Open	*Hattiesburg*	*Closed*
Florence	Open	Lake Charles	Open	Jackson	Open
Gadsen	Open	Leesville	Open	Kosciusko	Open
Huntsville	Open	Minden	Open	Laurel	Open
Jackson	Open	Monroe	Open	McComb	Open
Jasper	Open	Morgan City	Open	Meridian	Open
Mobile	Open	Natchitoches	Open	*Moss Point*	*Closed*
Montgomery	Open	New Iberia	Open	Natchez	Open
Opelika	Open	*New Orleans Bywater*	*Closed*	Newton	Open
Selma	Open	*New Orleans Downtown*	*Closed*	Olive Branch	Open
Talladega	Open	*New Orleans East*	*Closed*	Philadelphia	Open
Tuscaloosa	Open	*New Orleans West Bank*	*Closed*	Starkville	Open
		Opelousas	Open	Tupelo	Open
		Plaquemine	Open	Vicksburg	Open
		Ruston	Open		
		Shreveport	Open		
		Tallulah	Open		

Source: Compiled by the Congressional Research Service (CRS) from information provided by SSA.

PROCEDURES FOR DISPLACED PERSONS

A large number of residents of the areas affected by Hurricane Katrina have been temporarily relocated to other areas of their home states or other areas of the country. To accommodate these displaced persons, SSA has established satellite offices at the two largest evacuation shelters, the Astrodome in Houston, Texas and Kelly Air Force Base in San Antonio, Texas and is assigning employees to areas where the Federal Emergency Management Agency (FEMA) is providing services. The operating status of SSA field offices as of September 8, 2005, is summarized below and detailed in Table 2.

- Alabama: All 24 SSA field offices are open,
- Louisiana: 20 of 27 SSA field offices are open,
- Mississippi: 20 of 24 SSA field office are open.

TEMPORARY IDENTIFICATION PROCEDURES

It is expected that many of the residents of the areas affected by Hurricane Katrina may not have access to traditional forms of photo identification or their Social Security cards. In such cases, SSA staff are able to access the central computer system to verify identification based on Social Security Numbers and other information such as address and date of birth. According to SSA, once identified, individuals will be able to access their SSA benefits and receive a signed statement from SSA verifying their Social Security Number.

PROCEDURES FOR SSA EMPLOYEES

SSA is using existing employees displaced from their home offices to staff open field and satellite offices. SSA has established three special numbers for SSA employees to call to report their status and get instructions on where to report for work.

- Employees in the Atlanta Region (Alabama and Mississippi) should call (866) 616-4684
- Employees in the Dallas Region (Louisiana) should call (866) 635-8703
- Employees in the Louisiana Disability Determination Services offices should call (800) 587-3029

SSA ACTIVITIES AFTER SEPTEMBER 11, 2001

The attacks on the United States of September 11, 2001, were the last major national disaster that required an agency-wide response from SSA. After the terrorist attacks, the focus of SSA was in facilitating applications for disability and survivors benefits and in maintaining the continuity of agency operations in the affected areas of New York, Virginia, and Pennsylvania. Two years after the attacks, SSA reported that it had processed 5,629 benefit claims from 2,281 families and was paying ongoing benefits to 3,228 survivors and 498 persons disabled during the attacks.[8] To facilitate the processing of these benefits SSA took the following steps:[9]

- Assigned SSA staff to Family Assistance Centers located at Pier 94 in New York, Liberty State Park in New Jersey, and Arlington, VA. SSA staff were also available at the Flight 93 crash site in Shanksville, Pennsylvania and other locations including hospitals;
- Used the media and television spots to notify the public about SSA activities and procedures for filing for benefits;
- Relocated the New York Regional Office to an open New York Field Office and worked to reopen the closed New York offices quickly. All New York area offices, with the exception of one office located in the World Trade Center, were open by September 24, 2001;
- Worked with the United States Postal Service to ensure that benefits were paid when mail service was interrupted;
- Worked with the airlines, police and fire departments, trade unions, and affected companies to identify possible survivors and disability applicants;
- Processed claims for survivors benefits in the absence of a death certificate;
- Suspended the collection of overpayments from persons affected by the attacks.

As of this report, SSA is still considering various additional forms of response to Hurricane Katrina, including provisions for new disability and SSI applicants and the processing of survivors benefits.

SOURCES OF ADDITIONAL INFORMATION

Due to the dynamic nature of the situation in the states affected by Hurricane Katrina, it is likely that some of the information in this chapter may change as SSA refines its response to this crisis. Readers are encouraged to check with the Congressional Research Service (CRS) and the following additional sources for updated information.

Social Security Administration Hurricane Information [http://www.ssa.gov/emergency/]

Social Security Administration Hotlines (800) 772-1213 and (800) 325-0778 for TTY access

United States Postal Service Information [http://www.usps.com/communications/news/serviceupdates.htm?from=bannercommu nications&page=katrina]

Social Security Field Office Information: Alabama at [http://www.ssa.gov/atlanta/southeast/al/alabama.htm] Louisiana at [http://www.ssa.gov/dallas/state_la.html] Mississippi at [http://www.ssa.gov/atlanta/southeast/ms/mississippi.htm]

The Congressional Research Service (CRS)· Contact Scott Szymendera (x70014) or Kathleen Romig (x73742).

ENDNOTES

[1] Social Security Administration, *OASDI Beneficiaries by State and County, 2004,* (Washington: GPO, 2005), pp. 6, 39, 48.

[2] Social Security Administration, *SSI Recipients by State and County, 2004,* (Washington: GPO, 2005), pp. 3, 39, 50.

[3] For more information on Social Security programs see CRS Report 94-27, *Social Security: Brief Facts and Statistics,* by Gary Sidor; and CRS Report RL32279, *Primer on Disability Benefits: Social Security Disability Insurance (SSDI) and Supplemental Security Income (SSI)*, by April Grady and Julie M. Whittaker.

[4] Approximately 83% of all Social Security payments nationwide are currently paid through direct deposit, more information will be provided later in this report for the states affected by Hurricane Katrina.

[5] See CRS Report RS22245, *Postal Service for Katrina Survivors*, by Nye Stevens.

[6] For updated information, see the United States Postal Service website at [http://www.usps.com/ communications/news/serviceupdates.htm?from=bannercom munications&page=katrina#].

[7] Approximately 43% of SSI beneficiaries in Alabama, 40% in Louisiana, and 47% in Mississippi, receive their benefits through direct deposit. Data on direct deposit rates for retirement, disability, and survivors benefits can be found on SSA's website at [http://www.ssa.gov/deposit/T2StateSum_a.htm]. Data on direct deposit rates for SSI benefits can be found on SSA's website at [http://www.ssa.gov/deposit/T16State Sum_a.htm].

[8] Data taken from SSA Press Release available on SSA's website at [http://www.ssa.gov/pressoffice/factsheets/Sept11Anniv-2003.htm].

[9] *Social Security's Response to the September 11 Attacks: Hearing Before the House Committee on Ways and Means, Subcommittee on Social Security,* 107th Congress, (2001), (statement of Larry Massanari, Acting Commissioner for Social Security).

In: Federal Disaster Programs and Hurricane Katrina
Editor: Douglas D. Syzerhans, pp. 7-9
ISBN 1-59454-889-7

Chapter 2

HURRICANE KATRINA: FISHING AND AQUACULTURE INDUSTRIES — DAMAGE AND RECOVERY

Eugene H. Buck

SUMMARY

The Gulf Coast where Hurricane Katrina struck is an especially important center of commercial and recreational fishing, producing 10% of the shrimp and 40% of the oysters consumed in the United States. Many areas have been closed to fishing because of pollution-related contamination concerns. In addition, inland areas account for much of the U.S. farmed catfish production. This report summarizes damage assessments and recovery efforts, with initial reports primarily anecdotal until more accurate assessments become available. This report will be updated as warranted to incorporate new information.

SHRIMP

Commercial shrimpers fishing out of or delivering to Alabama, Mississippi, and Louisiana ports account for almost half of all U.S. shrimp production. Katrina has destroyed or severely damaged shrimp boats and shrimp processing and storage facilities throughout this area during this, the peak harvesting season. An uncounted number of shrimpers may have drowned trying to ride out the storm aboard their vessels, but information is not yet available. Unprocessed, rotting shrimp at damaged processing plants must be disposed. Even prior to Katrina, this segment of the U.S. fishing industry had been declining due to competition from less-expensive foreign imports and among domestic harvesters, since domestic capacity is much greater than necessary to efficiently harvest the resource. In addition, since shrimp trawling is very fuel consumptive, increasing fuel costs make shrimp trawling more uneconomical. Additional impediments to shrimping are the underwater obstacles that foul and damage shrimp trawls; hurricane debris will provide many new obstacles (i.e., "hangs"), and Katrina's storm surges may have moved former obstacles to new, uncharted positions.

Oysters

With the decline of oyster harvest from the Mid-Atlantic region, the Gulf of Mexico has been supplying most of the recent domestic oyster harvest. Oyster beds and oyster vessels along the Gulf Coast were extensively damaged, if not totally destroyed by Katrina. Because of extensive hurricane-related pollution and related contamination concerns, any remaining oysters will not be harvestable for an undetermined period.

Catfish Aquaculture

No reports have been received yet on the status of the farmed catfish industry.

Recreational Fishing

Damage to small boats and charter craft has been extensive; however, information is still sketchy on how this sector may have been affected.

Fishery Management

The National Marine Fisheries Service (NMFS) has many employees and contractors in the area damaged by Katrina. As of September 2, 2005, NMFS had made contact with 112 of 132 employees and contractors in the affected area. The NMFS facility at Pascagoula, MS, sustained significant damage.[1] The Gulf of Mexico Fishery Management Council meeting originally scheduled for September 12-16 in New Orleans has been postponed until October and moved to St. Petersburg, FL.

Seafood Consumers

While certain fish and shellfish from the Gulf may disappear from the market, extensive domestic and imported seafood alternatives remain. There could be some increase in price as retailers adjust to different products and suppliers. The price of oysters may be more affected than some other seafood products, because of less opportunity for substitution of similar items.

Disaster Assistance

Fishery disaster assistance is provided primarily through two authorities — §312(a) of the Magnuson-Stevens Fishery Conservation and Management Act (16 U.S.C. §1861a(a)) and

§308 of the Interjurisdictional Fisheries Act (16 U.S.C. §4107). These NMFS programs are further detailed at [http://sero.nmfs.noaa. gov/grants/fda.htm] and [http://www.nmfs.noaa. gov/mb/financial_services/disaster.htm].

Capacity Reduction

Distress in the commercial shrimp industry presents an opportunity for a possible capacity reduction effort to remove vessels and licenses permanently from the fleet. Such efforts could provide both compensation for damages for those who decide to sell their licenses and vessels as well as reduction in competition to those who may decide to resume shrimping. A summary of NMFS capacity reduction programs can be found at [http://www.nmfs.noaa.gov/mb/financial_services/buyback. htm].

Endnotes

[1] Photographs of damage were available at [http://www.nmfs.noaa.gov/pascagoula.htm] on Sept. 7, 2005.

In: Federal Disaster Programs and Hurricane Katrina
Editor: Douglas D. Syzerhans, pp. 11-17
ISBN 1-59454-889-7

Chapter 3

NEW ORLEANS LEVEES AND FLOODWALLS: HURRICANE DAMAGE PROTECTION

Nicole T. Carter

SUMMARY

Hurricane Katrina's storm surge breached floodwalls and levees surrounding New Orleans, causing widespread inundation and significant damage and hampering rescue and recovery efforts. Flooding from precipitation and storm surges flowing over levees was anticipated because of the hurricane's intensity; however, structural failure of the floodwalls and consequent flooding were uncertain. The immediate engineering and the underlying causes of the breaches are the subject of speculation, and likely will be the subject of investigations and congressional oversight. The breaches occurred at the *Lake Pontchartrain and Vicinity Project* being constructed by the U.S. Army Corps of Engineers and maintained by local levee districts. Those observers questioning why infrastructure providing a greater level of hurricane protection was not available are countered by those arguing that structural protections carry their own risks. This report will be updated as needed to track significant developments.

Flooding in New Orleans by Hurricane Katrina was predicted, but the extent of the inundation was uncertain. Because of the topography of the city — much of it below sea-level and lacking natural drainage — precipitation often causes local flooding which is controlled by a network of canals and pumps. Flooding as the result of overtopping of levees and floodwalls was predicted for a storm stronger than a fast-moving Category 3 hurricane; that is, it was expected that some water would flow over the levees and floodwalls based on Katrina's forecasted strength. Although the city did not receive Katrina's strongest force head-on, the levee and floodwall infrastructure's capability to protect the city was exceeded.

Failure (often called a breach) of levees and floodwalls reportedly was a contingency not central to emergency planning and response.[1] Four breaches along three New Orleans canals occurred as the result of Katrina. Failure is an inherent risk of any flood control structure, and can result in catastrophic damage. Breaches represent a structural failure, which is distinct from overtopping of levees and floodwalls which happens when their design is exceeded. The causes of the multiple breaches in New Orleans are the subject of speculation, and likely will be the subject of investigations and congressional interest.

This report focuses on the city's levee and floodwall infrastructure; it does not cover emergency planning, mitigation, response, and recovery efforts or other parts of the nation affected by Hurricane Katrina. This report is divided into three sections. The first provides a primer on the levee and floodwall system surrounding New Orleans, with particular attention to the federal projects and the challenges of protecting the city. The second section focuses on the floodwalls where the breaches occurred: the *Lake Pontchartrain and Vicinity Project.* The third section discusses providing coastal Louisiana with Category 4 and 5 hurricane protection.

New Orleans Storm and Flood Damage Reduction Infrastructure

New Orleans faces flooding threats from the Mississippi River, coastal storms, and intense precipitation; the system of levees and floodwalls around the city is designed to provide a certain level of protection from these threats.[2] The complimentary system of pumps and canals is designed to remove water trapped in the city. The storm damage reduction infrastructure around New Orleans consists of levees and floodwalls, and represents a combination of federal and local investments and responsibilities. Levees are broad, earthen structures, while floodwalls are concrete and steel walls, built atop a levee or in place of a levee. Floodwalls are often used in urban areas because they require less land than levees. The floodwalls in New Orleans generally are 6 to 10 feet tall and a foot wide at the top and two feet wide at the base, and they stand on earthen foundations.

Federal Projects

Most of the nation's flood and storm damage reduction infrastructure is maintained by local governments and local levee districts; some of the infrastructure is locally built, while other projects are built by the federal government. The principal federal agency responsible for constructing flood, storm, and shore protection infrastructure is the U.S. Army Corps of Engineers (Corps).[3]

Congress has authorized the Corps to construct five hurricane protection projects for coastal Louisiana; the two most relevant to New Orleans are the *Lake Pontchartrain and Vicinity Hurricane Protection Project* and the *West Bank and Vicinity Hurricane Protection Project*. In 1999, the House Committee on Transportation and Infrastructure authorized, via a Committee Resolution, a *Hurricane Protection Louisiana* study to investigate providing Category 4 and 5 hurricane protection for coastal Louisiana that would encompass all five of the previously-authorized projects.[4] Congress also authorized the *Southeast Louisiana Urban Flood Control Project* for pumping and canal improvements for Orleans, Jefferson, and St. Tammany parishes.

In addition, Congress has authorized a study of restoring and protecting coastal Louisiana's wetlands, which can buffer surges produced by some storms. Significant acreage of coastal Louisiana's wetlands have been converted to open water. This conversion is the result of many factors; some are human-induced, while others are natural processes. The most commonly cited human factors are (a) existing flood control and navigation improvements of the Mississippi River, which have reduced the natural processes of sediment nourishment of the coast, and (b) wetlands fragmentation and erosion from the oil and gas production infrastructure along Louisiana's coast, including the construction of numerous canals. The pending House and Senate versions of a Water Resources Development Act (WRDA) of 2005 (H.R. 2864 and S. 728) would authorize an initial $1.1 billion in efforts to restore these wetlands. For more information on coastal Louisiana wetlands loss and restoration, see CRS Report RL32673, *Coastal Louisiana: Attempting to Restore an Ecosystem,* by Jeffrey Zinn. For more information on WRDA legislation, see CRS Issue Brief IB10133, *Water Resources Development Act (WRDA): Army Corps of Engineers Authorization Issues in the 109th Congress*, coordinated by Nicole T. Carter.

Challenge of New Orleans Hurricane Protection

Protecting New Orleans has been an increasingly difficult task. First, the city has become increasingly vulnerable. Land in the city has subsided; barrier islands and wetlands buffering coastal Louisiana have been disappearing; and sea levels have risen. These factors have contributed to speculation that the existing storm damage protection infrastructure was not providing Category 3 protection. According to the project justification sheet included in the Administration's Corps FY2006 budget request, "the project was initially designed in the 1960s, and a reanalysis was performed for part of the project in the mid-1980s. Continuing coastal land loss and settlement of land in the project may have impacted the ability of the project to withstand the design storm."[5] Second, the greater New Orleans metropolitan area has been experiencing population and rapid economic growth resulting in greater numbers of people and more infrastructure and investments exposed to storm threats. The challenge of protecting New Orleans could become even greater. According to some scientists, higher sea surface temperatures may result in increased hurricane intensity.[6] Climate change concerns and other factors have raised questions about whether both estimates of the likelihood of hurricanes of various strengths and past decisions based on these estimates need to be re-evaluated.

LAKE PONTCHARTRAIN AND VICINITY PROJECT

Breaches and Flooding

Some parts of New Orleans were flooded by Hurricane Katrina's precipitation and overtopping of levees, but the major source of flood waters appears to have been the floodwall breaches of the Lake Pontchartrain and Vicinity Project. The immediate and underlying causes of these breaches are the subject of much speculation, ranging from poor or exceeded design, to faulty construction, to poor maintenance, to slow construction due to limited Corps appropriations.

According to the Corps,[7] the breaches were at floodwalls along three canals. The 17th Street Canal breach was at a levee-floodwall combination; the breach reached 450 feet in length. According to the Corps, "it's believed that the force of the water overtopped the floodwall and scoured the structure from behind and then moved the levee wall horizontally." The Industrial Canal floodwall had two breaches — a 100-foot breach, and a 500-foot breach.

The London Street floodwall had a 300-foot breach. The Corps continues to work on repairing the breaches. The Corps noted that other levees and floodwalls were being monitored for potential failure. One reason to fear additional failures is that levees and floodwalls are not designed for extended retention of water, but are instead designed for short-term storm events.

Removing flood waters from the city presents multiple challenges. First, repair of the 17th Street Canal is important because the canal is the main route for moving water out of the city once pumps are running. Second, most of the city's pumping facilities are not operational. Estimates for removing flood waters range from 36 days to more than 2 months. The city's saturated soil may delay subsequent rebuilding.

Project History and Appropriations

The Lake Pontchartrain and Vicinity Project was authorized by Congress originally in the Flood Control Act of 1965, with subsequent modifications in other legislation. The project, which remains under construction, is intended to reduce hurricane damages from surges entering Lake Pontchartrain from Lake Borgne through natural tidal passes. The total cost for the Lake Pontchartrain and Vicinity project is \$738 million, with the federal responsibility \$528 million. Federal allocations reached \$458 million (87% of the federal responsibility) with the amount appropriated for FY2005. A Corps project fact sheet stated that the project's FY2005 appropriations would be "insufficient to fund new construction contracts," and that the Corps had the capability to use \$20 million.[8] The fact sheet also stated:

> In Orleans Parish, two major pump stations are threatened by hurricane storm surges. Major contracts need to be awarded to provide fronting protection for them. Also, several levees have settled and need to be raised to provide the design protection. The current funding shortfalls in fiscal year 2005 and fiscal year 2006 will prevent the Corps from addressing these pressing needs.

Table 1. Requests and Appropriations for Selected New Orleans Projects and Corps Civil Works

Fiscal Year	Lake Pontchartrain (in thousands)		West Bank and Vicinity (in thousands)		Southeast Louisiana (in thousands)		*Total Corps Civil Works (in billions)*	
	Req.	App.	Req.	App.	Req.	App.	*Req.*	*App.*
2005	3,937	5,719	37,000	30,000	30,000	36,500	*4.12*	*4.71*
2004	3,000	5,500	35,000	28,500	16,500	34,400	*4.19*	*4.57*
2003	4,900	7,000	5,000	9,000	20,083	50,000	*4.29*	*4.63*
2002	7,500	14,250	12,000	12,500	51,908	60,000	*3.90*	*4.49*
2001	3,100	10,000	8,065	8,065	47,260	69,000	*4.06*	*4.54*
2000	11,887	16,887	7,000	15,070	47,066	47,066	*3.91*	*4.14*
1999	5,676	16,000	3,936	7,000	15,278	75,000	*3.24*	*3.96*
1998	6,448	22,920	2,385	2,385	6,440	47,000	*3.70*	*4.06*
1997	4,025	17,025	0	1,500	10,000	17,500	*3.29*	*3.50*
1996	7,848	13,348	NA	NA	0	2,000	*3.34*	*3.20*

Source: Compiled by CRS from Energy and Water Appropriations conference reports.

Funding for the Lake Pontchartrain project has declined from 1996 through 2005. During that period, the Clinton and Bush Administrations requested $58 million for the project; Congress appropriated approximately $70 million more than requested. A funding trend for the Corps' two other New Orleans projects, however, is less evident. The funding history for the Lake Pontchartrain hurricane protection, West Bank hurricane protection, and Southeast Louisiana Urban Flood Control projects and for the total Corps civil works program is provided in Table 1. Although the Corps civil works appropriations have increased in the last decade, competition for funds among projects has been tight, with many projects that historically received appropriations not being included in recent budget requests. Federal investments in water resources infrastructure remains considerably lower than levels of the 1950s and 1960s,[9] and the Corps has warned of problems associated with aging infrastructure and a backlog of construction activities.

CATEGORY 4 AND 5 PROTECTION

In August 2002, the Corps completed a reconnaissance study of whether to strengthen coastal Louisiana's hurricane damage reduction projects, including the New Orleans projects, to protect against Category 4 and 5 storms. If implemented, coastal Louisiana would have the only Category 5 protection system in the country. The next step in the Corps' project development process is a feasibility study.[10] The feasibility study for this project was estimated to cost $8 million and take at least five years; construction would take another 10 to 20 years and $2.5 billion.[11] A congressionally-added FY2005 appropriation of $100,000 for the feasibility study was to be used for completing a project management plan and executing the cost share agreement with the local sponsor of the study. The study was not included in the President's FY2006 request. According to a Corps document, $500,000 for FY2006 is required to initiate work on the feasibility study.[12] The House report (H.Rept. 109-86) for the pending Energy and Water Development Appropriations Act for FY2006 (H.R. 2419) did not include funds for this study; the Senate report (S.Rept. 109-84) included $250,000.

As part of a feasibility study, the Corps analyzes the national benefits and costs of a storm damage reduction project. How these evaluations are performed and the costs and benefits that are included has long been debated. For example, structures in New Orleans area generally were not built to withstand Category 5 winds; therefore, providing Category 5 protection from storm surges will not protect the city and its structures from a Category 5 hurricane's winds and other threats. These analyses also take into account the likelihood of various storm events happening.

Hurricane Katrina has resulted in some questioning why a Category 4 or 5 hurricane storm damage system was not already in place for New Orleans, and whether it should be part of the rebuilding effort. Others are supporting more emphasis on restoration of coastal wetlands to improve storm damage protection. Others are raising concerns about the extent of rebuilding that should take place considering the city's hurricane and flooding vulnerability. They are concerned that higher levees and floodwalls may provide a false sense of security and continued risk of catastrophic failure. Decisions on what type and level of hurricane protection to provide New Orleans and other coastal areas in the future, as well as factors contributing to Katrina's damages (Category 3 infrastructure, and levee and floodwall construction, maintenance, and appropriations) likely will be the subject of congressional oversight.

ENDNOTES

[1] S. Shane and E. Lipton, "Government Saw Flood Risk but Not Levee Failure," *The New York Times*, Sept. 2. 2005.

[2] For a map of this system produced by the *New Orleans Times-Picayune*, see [http://www.nola.com/hurricane/popup/nolalevees_jpg.html], visited on Sept. 6, 2005.

[3] For more on the Corps' water resources activities, see CRS Report RS20866, *The Civil Works Program of the Army Corps of Engineers: A Primer*, by Nicole T. Carter and Betsy A. Cody.

[4] Storm Categories are based on the Saffir-Simpson Hurricane Scale of 1-5, with Category 4 storms having 131-155 mile per hour (mph) winds and a 13- to 18-foot storm surge and Category 5 storms having winds exceeding 155 mph and surges above 18 feet.

[5] U.S. Army Corps of Engineers, *Fiscal Year 2006, Mississippi Valley Division*, visited Sept. 3, 2005, available at [http://www.usace.army.mil/inet/functions/cw/cecwb/just_states06/mvd.pdf].

[6] For example, see Intergovernmental Panel on Climate Change, *Climate Change 2001: Working Group II: Impacts, Adaptation and Vulnerability*, visited on Sept. 3, 2005, available at [http://www.grida.no/climate/ipcc_tar/wg2/029.htm], or T.R. Knutson and R.E. Tuleya, "Impact of CO2-Induced Warming on Simulated Hurricane Intensity and Precipitation: Sensitivity to the Choice of Climate Model and Convective Parameterization," *Journal of Climate*, 15 Sept. 2004, 17 (18), available at [http://www.gfdl.noaa.gov/reference/bibliography/2004/tk0401.pdf].

[7] This Sept. 1, 2005 document is available at [http://www.mvd.usace.army.mil/hurricane/news/11_pm_1_sept_corps_response_to_katrina_ recovery_efforts.doc], visited on Sept. 3, 2005.

[8] U.S. Army Corps of Engineers, *Project Fact Sheet, Lake Pontchartrain, LA. and Vicinity Hurricane Protection Project* (May 23, 2005) available at [http://www.mvn.usace.army.mil/pao/response/HURPROJ.asp?prj=lkpon1], visited Sept. 3, 2005.

[9] For example, annual appropriations for the Corps' construction account fell from a $4 billion average in the mid-1960s (in 1999 dollars) to a $1.5 billion average for 1996 through 2005.

[10] For information on the Corps project development process, see CRS Report RL32064, *Army Corps of Engineers Water Resources Activities: Authorization and Appropriations*, by Nicole T. Carter and H. Steven Hughes.

[11] U.S. Army Corps of Engineers, New Orleans District, *Riverside, September-October 2004*, visited on Sept. 3, 2005, available at [http://www.mvn.usace.army.mil/pao/Riverside/Sept-Oct_04_Riv.pdf]; and A.C. Revkin, and C. Drew, "Engineers' warnings and pleas for money went unheeded,"*The New York Times*, Sept. 2, 2005.

[12] The March 2005 fact sheet, visited on Sept. 3, 2005, is available at [http://www.mvn.usace.army.mil/pd/Funding_Programs/Current/GI_HurrProt_FY06.pdf].

In: Federal Disaster Programs and Hurricane Katrina
Editor: Douglas D. Syzerhans, pp. 19-23
ISBN 1-59454-889-7

Chapter 4

DISASTER EVACUATION AND DISPLACEMENT POLICY: ISSUES FOR CONGRESS

Keith Bea

SUMMARY

The devastation caused by Hurricane Katrina reaches beyond the borders of the states directly affected by the wind, rain, and floods. Before the storm reached the coast, thousands of residents of Louisiana and Mississippi evacuated to other states, including Texas and Oklahoma. Many people, for a variety of reasons, chose to disregard the mandatory evacuation orders issued by state and local officials. In general, evacuation policy is set and enforced by state and local officials. Federal policy provides for various aspects of civilian evacuation. As Members of Congress explore the challenges and losses in the states affected directly or indirectly by Hurricane Katrina, they may be called upon to consider federal policy options to more fully integrate federal and state authorities.

Using the authority set out in state laws and local ordinances, state and local officials may suggest or require the evacuation of residents from homes and communities before certain catastrophes occur.[1] Threats of pending natural disasters such as hurricanes or floods, warnings of the movement of airborne hazardous material due to transportation accidents, or the recognition of unstable conditions at nuclear power plants may provide officials a short window of opportunity to save thousands of lives.

To a limited extent, federal statutes authorize agency heads to use federal resources to assist in the evacuation of civilians. Among those authorities, and arguably pertinent to the situation in New Orleans even before Hurricane Katrina made landfall, is the provision in the Robert T. Stafford Disaster Relief and Emergency Assistance Act for the President to direct the Secretary of Defense to use resources to perform "emergency work which is made necessary by such incident and which is essential for the preservation of life and property."[2]

The complications and suffering associated with Hurricane Katrina demonstrate that the evacuation and the relocation of tens of thousands of residents pose enormous challenges for public officials after catastrophes occur. Due to limitations involving medical, income, or transportation needs, or for other reasons, many did not or could not heed the warnings. The flooding of New Orleans and the widespread destruction of infrastructure in Mississippi and Louisiana caused by Hurricane Katrina may result in tens of thousands of persons remaining homeless for many months, and possibly in the relocation of entire communities. The issue is whether federal policy regarding the evacuation of civilians and their resettlement, or the implementation of that policy, is appropriate, and whether the policy might be better integrated with state authorities.

EVACUATION POLICY

In general, federal policy defers to the states to enact laws pertinent to evacuation, and local officials generally work with state officials to enforce those laws. For example, the *National Response Plan* (NRP), administered by the Federal Emergency Management Agency (FEMA) in the Department of Homeland Security (DHS), sets forth the roles and responsibilities of federal and certain non-federal entities after catastrophes overwhelm state or local governments. The NRP includes the following in a list of responsibilities of local chief executive officers:

> Dependent upon state and local law, [the local executive] has extraordinary powers to suspend local laws and ordinances, such as to establish a curfew, direct evacuations, and, in coordination with the local health authority, to order a quarantine.[3]

Federal officials facilitate evacuation procedures when the NRP is invoked, take the lead on coordinating necessary decisions, support search and rescue efforts, and provide technical assistance.[4] In the event of the most catastrophic incidents (such as after Hurricane Katrina devastated the Gulf Coast on August 29, 2005), "the federal government or other national entities" are expected to provide public health, medical, and mental health support at casualty evacuation points and refugee shelters."[5]

Federal law contains provisions related to the evacuation of civilians and planning requirements within the United States, under specified circumstances. General authority is also given to federal agencies for emergency preparedness activities that include evacuation planning. These authorities are cited in Table 1.

DISPLACEMENT AND SHELTERING

One facet of evacuation is the removal of persons from a stricken or threatened area. Another is the resettlement of the victims, or potential victims, and the provision of shelter and resources in the areas to which those residents are moved, known as the host areas. The NRP vests responsibility for the coordination of mass care for disaster victims in FEMA and the American Red Cross. With the support of other federal agencies, charitable organizations, the private sector, and individuals, victims of catastrophes may find shelter outside the stricken area for a temporary period of time.[6]

Table 1. Federal Statutory Provisions on Evacuation Policy

Summary	Citation
Federal employees and their dependents may receive assistance if they must be evacuated.	5 U.S.C. 5709, 5725, 5922, 5923
The role of FEMA includes evacuating disaster victims.	6 U.S.C. 317
National Construction Safety Teams must evaluate technical aspects of evacuation procedures and recommend research.	15 U.S.C. 7301, 7307-7308
Chief of Engineers may use funds to evacuate persons in a flood wall project area if the cost of the project can be substantially reduced.	33 U.S.C. 701i
Emergency preparedness activities include non-military civilian evacuation and evacuation of personnel during hazards.	42 U.S.C. 5195a
Computer models for evacuation must be periodically evaluated and improved.	42 U.S.C. 7403(f)(2)(C)
Temporary housing and evacuation of threatened persons are to be included in the scope of hazardous substance removal.	42 U.S.C. 9601(23)
Emergency plans completed by local emergency planning committees (LEPCs) must include evacuation plans.	42 U.S.C. 11003
Owners of facilities where a hazardous chemical release occurs must provide information on precautions to be taken, including evacuation.	42 U.S.C. 11004(b)(2)
Secretary of Transportation must establish incident response plans for facilities and vessels that include evacuation procedures.	46 U.S.C. 70104(b)
Congressional finding made that private and public sector emergency preparedness activities should include an evacuation plan.	P.L. 108-458, §7305, 118 Stat. 3848
Evacuation routes may be included as components of the National Highway System under the high priority corridor designations.	H.R. 3 (109th Congress) Sec. 1304 (a) Signed by President George W. Bush on August 10, 2005
The Secretary of the Department of Transportation (DOT) and the Secretary of Homeland Security in coordination with the Gulf Coast States and contiguous states, must review and assess federal and state evacuation plans for catastrophic hurricanes impacting the Gulf Coast Region and report, by October 1, 2006, findings and recommendations to Congress.	H.R. 3 (109th Congress) Sec. 10204
The National Science Foundation is to produce a public transportation security study of public systems' ability to accommodate the emergency evacuation, egress, or ingress from urban areas with populations over one million.	H.R. 3 (109th Congress) Sec. 3046
Emergency evacuation studies are a required DOT activity under the deployment of the transportation model known as the "Transportation Analysis Simulation System."	H.R. 3 (109th Congress) Sec. 5512(b) (4)

Source: CRS compilation based on database searches conducted by Thomas P. Carr of the Government and Finance Division and contributions by Robert Kirk of the Resources, Science, and Industry Division, both of CRS.

Note: Table 1 excludes statutory provisions related to military personnel, criminal offenses, foreign nations and international relations, liability, and payment of costs.

The individual assistance and temporary housing provisions of the Stafford Act authorize federal agencies to provide alternative housing for disaster victims.[7] FEMA administers this program that provides financial assistance for alternative accommodations, or direct assistance through modular housing moved to a temporary housing site. In order to make use of this authority, vacant alternative housing units (apartments, motels or hotels, recreational vehicles, modular units) must be found, or sites must be developed for the mobile homes brought on-site by FEMA.

Issues for Congressional Consideration

- The President, at the request of the Governor of a state, is authorized to direct the Secretary of Defense to use resources for emergency work to save lives. The Governors of the states are authorized to direct the evacuation of civilians under the emergency statutory authorities. In the wake of Hurricane Katrina Congress may assess how effectively federal authority pertinent to evacuations is implemented by Administration officials.
- Many catastrophes, natural and human-caused, cannot be anticipated; nor can threats or warnings be provided. Congress may review statutory means to expedite the delivery of federal resources and personnel to disaster stricken areas in order to evacuate civilians and provide necessary medical treatment.
- The Catastrophic Incident Annex of the NRP includes the assumption that large-scale evacuations, whether spontaneous or directed pursuant to authorities, may be more likely to occur after attacks than natural disasters, and that the maintenance of public health will "aggravate attempts to implement a coordinated evacuation management strategy."[8] In light of concerns raised by some that the NRP, and the focus of DHS, is skewed toward concerns about terrorist attacks, Congress may review whether this assumption adequately addresses the complications evident in the evacuation of New Orleans before and after Hurricane Katrina struck.
- Persons in the affected population who struggle with disabilities, be they mental or health, need particular attention in evacuations. The National Organization on Disability has developed a guide for the consideration of the needs of disabled persons, particularly the evacuation of those individuals from their communities.[9] Congress could consider whether standing state authorities adequately address the needs of the disabled in evacuation circumstances.
- The Stafford Act stipulates that relief and assistance be provided "without discrimination on the grounds of...economic status."[10] Through authorities cited in Table 1 of this report, FEMA has responsibility to provide for the evacuation of disaster victims,[11] and provide for evacuation as part of federal emergency preparedness efforts.[12] Congress may assess whether existing evacuation plans and procedures comport with the requirements of the Stafford Act.
- Temporary housing is authorized to be provided generally for 18 months after a disaster. The detonation of a nuclear device in a metropolitan area, or a major incident at a nuclear power plant, could result in long-term consequences that might render areas uninhabitable for years. The destruction associated with Hurricane Katrina may force thousands of families to seek alternative shelter for many months,

if not years. Congress may review existing statutory authorities given this longer term issue.

- Federal law does not authorize assistance for states indirectly affected by catastrophes. States that house tens of thousands of disaster victims for many months will face unexpected service delivery burdens. Congress might examine federal policy options to provide assistance to host states.

ENDNOTES

[1] State laws generally authorize the Governor to order and enforce the evacuation of residents under emergency situations. See CRS Report RL32287, *Emergency Management and Homeland Security Statutory Authorities in the States, District of Columbia, and Insular Areas: a Summary*, by Keith Bea, Government and Finance Division, CRS, and L. Cheryl Runyon and Kae M. Warnock, consultants, p. 4.

[2] 42 U.S.C. 5170b(c).

[3] U.S. Department of Homeland Security, *The National Response Plan* (Washington: 2004), p.

8. Hereafter cited as NRP. Similar text is used to describe the role of tribal chief executive officers. See also p. 53 of the NRP (initial actions taken by first responders and local government authorities including evacuations) and p. 54 (response actions that may involve federal assistance include law enforcement and evacuations).

[4] For example, the Department of Transportation (DOT) Regional Emergency Transportation Coordinator (RETCO) coordinates with state and local authorities on "issues such as movement restrictions, critical facilities closures, and evacuations;" establishes resource priorities, and provides technical assistance. Ibid., p. ESF#1-3, 4. The Department of Defense coordinates with DOT to evacuate patients, at the request of the Department of Health and Human Services. Ibid., p. ESF#8-4. See also 44 CFR 352.26, which provides that federal agencies "may be called upon" to provide reception centers or shelters for evacuees.

[5] NRP, p. CAT-2.

[6] "Emergency shelter includes the use of pre-identified shelter sites in existing structures...and use of similar facilities outside the incident area, should evacuation be necessary." Ibid., p. ESF#6-2.

[7] 42 U.S.C. 5174.

[8] Ibid., p. CAT-3.

[9] National Organization on Disability, *Emergency Preparedness Initiative: Guide on the Special Needs of People with Disabilities for Emergency Managers, Planners & Responders*, available at [http://nod.org/index.cfm?fuseaction=Page.viewPage&pageId=11], visited Aug. 31, 2005.

[10] 42 U.S.C. 5151(b).

[11] 6 U.S.C. 317.

[12] 42 U.S.C. 5195a.

In: Federal Disaster Programs and Hurricane Katrina
Editor: Douglas D. Syzerhans, pp. 25-28
ISBN 1-59454-889-7

Chapter 5

OIL AND GAS: SUPPLY ISSUES AFTER KATRINA

Robert L. Bamberger and Lawrence Kumins

SUMMARY

Hurricane Katrina made landfall on August 29, 2005, leaving behind considerable devastation. In anticipation of the hurricane's landfall, workers were evacuated from oil rigs in the Gulf region. Some onshore refineries were shut down in advance of the storm; others are also down now because of the widespread interruption of electric power. Assessment of damage to oil and gas production rigs, as well as refineries, has begun. It is uncertain when electricity will be restored. Until more is known, the outlook for oil and gas is unclear. In the meantime, spot and futures prices for gasoline and middle distillates are rising sharply. This report will be updated.

Hurricane Katrina has shut down oil and gas production from the Outer Continental Shelf in the Gulf of Mexico, the source for 25% of U.S. crude oil production. Additionally, there are a number of oil refineries shut down along the Gulf Coast in Louisiana and Mississippi that provide a significant share of the nation's refined petroleum products, much of which is transported by pipeline for the East Coast and upper Midwest. Important to policymakers will be information on the extent of the damage from Hurricane Katrina to production facilities, refineries, and transport facilities and how soon they can be restored to operation. Most of the information cited in this report has come from the trade press and the Energy Information Administration.

The policy option most mentioned during the first hours after the hurricane was the Strategic Petroleum Reserve (SPR). On August 31, Secretary of Energy Bodman indicated that the Bush Administration was authorizing releases of SPR crude oil. Operating refineries whose crude supply will be interrupted by hurricane damage can request a loan or swap of crude from the SPR. Roughly 5.4 million barrels of SPR oil was loaned to refiners in the aftermath of Hurricane Ivan in September 2004. However, as a policy tool to respond to the crisis, the SPR has limitations because a barrel of crude contributes to product supply only if there is refining capacity to turn the crude into gasoline or diesel fuel. Consequently, recovery from the hurricane's effects will depend upon resumption of production and refining operations in the Gulf, and the ability to transport petroleum products.[1]

FACTORS AFFECTING OIL AND NATURAL GAS SUPPLY

- *Outer Continental Shelf (OCS) Production.* Natural gas and petroleum production in the Western Gulf of Mexico is virtually halted. Eastern Gulf output is seriously curtailed. About 1.4 million barrels per day (mbd) of crude and 8.8 billion cubic feet per day (bcf/d) of gas output are currently offline.[2] This is the equivalent of 6.5% of U.S. petroleum consumption and 15% of natural gas demand.
- *Suspension of Natural Gas Deliveries at Henry Hub.* This is the delivery point, located in Louisiana, for consumers taking physical delivery of gas traded on the New York Mercantile Exchange (NYMEX), where such trading has been suspended. The Henry Hub gas contract is an important pricing benchmark widely used in setting prices nationwide, and the suspension of trading leaves the industry without guidance from this visible public auction market.
- *Lack of Alternative Gas Supplies.* OCS natural gas supply losses are not easily made up by imports because additional supply possibilities from Canada and from liquefied natural gas (LNG) are limited. There is no Strategic Petroleum Reserve (SPR) for gas. A short-term problem could worsen if the shortage continues into the heating season. The status of natural gas pipelines has yet to be determined.
- *Crude Oil Imports via Louisiana Offshore Oil Port (LOOP).* About 1 mbd of crude oil imports flow through this tanker offloading facility, which is now closed. This constitutes roughly 5% of petroleum consumption.[3] Other Gulf Coast ports through which oil imports flow are not operating; the shortfall in crude oil supply likely involves more than the LOOP and OCS production loss. It has been reported that damage to LOOP is minimal to the offshore facility and that it could resume operation once power is restored.
- *Pipeline Transport of Crude Oil and Refined Products.* Pipelines from the Gulf to the Midwest and East Coast have been impacted. The Colonial Pipeline — serving the whole east coast with refined products — is currently closed. While there has been no statement from the pipeline owners regarding the nature of the difficulties, lack of product input and electric power supply are likely considerations.

REFINERIES

Nine major facilities (and three small plants) in Louisiana and Mississippi with 2.1 mbd capacity are closed;[4] this constitutes 12.4% of total U.S. capacity, and is the equivalent of about 10% of total refined product consumption. (See Table 1.) This includes an estimated 1.2 mbd of gasoline (13% of consumption) and 0.4 mbd of distillate (10% of consumption). These refineries were shut down before the storm. Located in low-lying areas, they are currently underwater, precluding physical inspection. Inspection from aircraft suggests minimal damage, although there is no electric power available.

Table 1. Major Refineries Affected by Hurricane Katrina

Capacity (thousands of barrels per day)	Owner	Location	Status as of 8/31
493.5	ExxonMobil	Baton Rouge, LA	Briefly operating at reduced capacity.
187.0	ExxonMobil/PDVSA	Chalmette, LA	Shut. Water damage likely.
325.0	Chevron	Pascagoula, MS	Shut.
226.0	Motiva Enterprises[6]	Norco, LA	Shut.
235.0	Motiva Enterprises	Convent, LA	Shut.
247.0	ConocoPhillips	Belle Chase, LA	Shut. Water damage expected.
245.0	Marathon-Ashland	Garyville, LA	Shut.
120.0	Murphy Oil	Meraux, LA	Shut. Water damage expected.
185.0	Valero Energy	St. Charles, LA	Out 1-2 weeks. Water damage expected.

Source: Energy Information Administration, Office of Electricity Delivery and Energy Reliability. Hurricane Katrina Situation Report #12. August 31, 2005 (10:00 AM EDT).

Gasoline and Distillate Inventories

As of August 26, the Energy Information Administration (EIA) reports that U.S. commercial gasoline inventories were 194 million barrels, near the lower bound of the normal range. Minimum operating level is 170 million barrels, representing the point at which localized shortages occur because the distribution system cannot be drawn down further without hindering operations. At current consumption rates, this amounts to less than three days of available supply.

The situation with middle distillates (heating oil and diesel fuel) is somewhat similar, reflecting the normal seasonal build of stocks, now standing at 135 million barrels, the equivalent of nearly nine days of available supplies.

As a result of Katrina, wholesale gasoline prices on the NYMEX surged almost immediately, in anticipation of a shortage. On August 31, wholesale gasoline traded on the NYMEX reached $2.52 per gallon, a new high. Pump prices also reached a new high of $2.62 nationwide, according to the American Automobile Association. Distillate prices followed, albeit with a lower rate of increase.

Commercial Inventories of Crude Oil

Crude stocks were 321 million barrels on August 26, up from 285 million barrels during the same period last year. Minimum operating level for crude is 270 million barrels, suggesting that 51 million barrels might be available from existing inventories. This could be used to offset the shortfall of OCS production and imports, if it were located where it could be accessed by refiners seeking replacement crude. But it is not clear that logistics would support this, nor is it clear when the 2.1 mbd of shut refining capacity could be restored to utilize these inventories.

Strategic Petroleum Reserve

On August 31, Secretary of Energy Bodman indicated that the Administration was authorizing loans of crude oil from the Strategic Petroleum Reserve. Three refineries are reported to have requested SPR oil, only one of which has yet been identified. Citgo Petroleum Corporation has asked for 250,000-500,000 barrels for its refinery at Lake

Charles, Louisiana.[5] But as noted above, the SPR will be of limited help in this situation if there is insufficient refining capacity to turn its stocks of crude oil into gasoline or diesel fuel.

Endnotes

[1] For additional information, see CRS Issue Brief IB87050, *The Strategic Petroleum Reserve*, by Robert L. Bamberger.

[2] Energy Information Administration, *Special Report — Hurricane Katrina's Impact on U.S. Energy*, at [http://tonto.eia.doe.gov/oog/special/eia1_katrina.html], August 30, 2005.

[3] Ibid.

[4] Platts Oilgram Price Report, *Saudi Oil, SPR Loans Promised To Ease Shortages*, August 29, 2005. p 1.

[5] "White House to Release Oil From Reserves," Eileen Putnam, Associated Press.

[6] Motiva is a joint venture between Royal Dutch Shell and Saudi Aramco.

In: Federal Disaster Programs and Hurricane Katrina
Editor: Douglas D. Syzerhans, pp. 29-37
ISBN 1-59454-889-7

Chapter 6

FEDERAL STAFFORD ACT DISASTER ASSISTANCE: PRESIDENTIAL DECLARATIONS, ELIGIBLE ACTIVITIES, AND FUNDING

Keith Bea

SUMMARY

The Robert T. Stafford Disaster Relief and Emergency Assistance Act (the Stafford Act) authorizes the President to issue a major disaster declaration to speed a wide range of federal aid to states determined to be overwhelmed by hurricanes or other catastrophes. Financing for the aid is appropriated to the Disaster Relief Fund (DRF), administered by the Department of Homeland Security (DHS). Funds appropriated to the DRF remain available until expended (a "no-year" account). The Stafford Act authorizes temporary housing, grants for immediate needs of families and individuals, the repair of public infrastructure, emergency communications systems, and other forms of assistance.

Because the Stafford Act provides the President with permanent authority to direct federal aid to stricken states, Congress need not enact new legislation to meet immediate needs. Congress appropriated over $10 billion to the DRF in FY2005, largely in response to the four hurricanes that struck Florida in the fall of 2004. The appropriations legislation for FY2006 includes roughly $2 billion for the DRF in both the House and Senate versions of H.R. 2360 in conference at the time Hurricane Katrina struck. Congress can elect to consider supplemental appropriations should additional money be required to meet the requests for assistance.

OVERVIEW OF THE STAFFORD ACT

The Robert T. Stafford Disaster Relief and Emergency Assistance Act (the Stafford Act) authorizes the President to issue major disaster declarations that authorize federal agencies to provide assistance to states overwhelmed by disasters.[1] Through executive orders, the President has delegated to the Federal Emergency Management Agency (FEMA), within the Department of Homeland Security (DHS), responsibility for administering the major provisions of the Stafford Act. Assistance authorized by the statute is available to individuals, families, state and local governments, and certain nonprofit organizations.

Activities undertaken under authority of the Stafford Act are provided through funds appropriated to the Disaster Relief Fund (DRF). Federal assistance supported by DRF money is used by states, localities, and certain non-profit organizations to provide mass care, restore damaged or destroyed facilities, clear debris, and aid individuals and families with uninsured needs, among other activities. In calendar year 2004, President Bush issued 68 major disaster declarations; in calendar year 2005, 32 such declarations have been issued, including those for Florida, Louisiana, and Mississippi for Hurricane Katrina.[2]

Presidential Declarations

Under Stafford Act authority, five types of actions may be taken, summarized as follows.

- **Major disaster.** The President issues a major disaster declaration after receiving a request from the governor of the affected state.[3] Major disaster declarations may be issued after a natural catastrophe or, "regardless of cause, fire, flood or explosion."[4] A declaration authorizes DHS to administer various federal disaster assistance programs for victims of declared disasters. Each major disaster declaration specifies the type of incident covered, the time period covered, the types of disaster assistance available, the counties affected by the declaration, and the name of the federal coordinating officer.
- **Emergency**. The declaration process for emergencies is similar to that used for major disasters; the President may, however, issue an emergency declaration without a gubernatorial request if primary responsibility rests with the federal government.[5] An emergency declaration may be issued on "any occasion or instance" in which the President determines that federal assistance is required. Under an emergency declaration, the federal government funds and undertakes emergency response activities, debris removal, and individual assistance and housing programs. DRF expenditures for an emergency are limited to $5 million per declaration unless the President determines that there is a continuing need; Congress must be notified if the $5 million ceiling is breached.
- **Fire suppression.** The Secretary of DHS is authorized to provide fire suppression assistance to supplement the resources of communities when fires threaten such destruction as would warrant a major disaster declaration.
- **Defense emergency**. Upon request from the governor of an affected state, the President may authorize the Department of Defense (DOD) to carry out emergency work for a period not to exceed 10 days. DOD emergency work is limited to work essential for the preservation of life and property.
- **Pre-declaration activities**. When a situation threatens human health and safety, and a disaster is imminent but not yet declared, the Secretary of DHS may place agency employees on alert. DHS monitors the status of the situation, communicates with state emergency officials on potential assistance requirements, and deploys teams and resources to maximize the speed and effectiveness of the anticipated federal

response and, when necessary, performs preparedness and preliminary damage assessment activities.

In considering a gubernatorial request for disaster relief, the President evaluates a number of factors, including the cause of the catastrophe, damages, needs, certification by state officials that state and local governments will comply with cost sharing and other requirements, and official requests for assistance. FEMA has established thresholds which are considered by the President and DHS officials in the process of determining whether a major disaster is to be declared.[6] Neither the Stafford Act nor implementing regulations provide for a congressional role in the declaration process.[7]

Types of Assistance and Eligibility

FEMA has established three major categories of aid under the Stafford Act — individual and household, public, and hazard mitigation assistance. The persons and organizations eligible for assistance authorized by the Stafford Act may be summarized in the following fashion:

- **Individuals and households** — immediate temporary shelter, cash grants (maximum of approximately $25,000, adjusted for inflation) for uninsured emergency personal needs, temporary housing assistance (rental and mortgage payments) generally for 18 months, home repair grants, unemployment assistance due to the disaster, debris removal from private property when deemed in the public interest, emergency food supplies, legal aid for low-income individuals, and crisis counseling;
- **State, tribal, and local governments and certain private nonprofit organizations**[8] — repair, reconstruction, or replacement of infrastructure and recreational facilities; emergency protective measures, emergency communications and transportation systems; and loans to replace lost revenue or meet federal cost-sharing requirements; and,
- **State governments** — hazard mitigation assistance to reduce future disaster losses.

Disaster Relief Funding

Congress appropriates money to the Disaster Relief Fund (DRF) to ensure that the foregoing federal assistance is available to help individuals and communities stricken by severe disasters. Funds appropriated to the DRF remain available until expended.[9]

Appropriations to the DRF generally evoke little controversy. Supplemental appropriations legislation is generally required each fiscal year to meet the urgent needs of particularly catastrophic disasters. Questions have been raised in the past concerning the increased cost of federal disaster assistance authorized by the Stafford Act as the categories of aid and eligibility for federal disaster assistance have expanded.

Congress has previously explored the issue of rising federal disaster assistance costs and reliance upon supplemental appropriations.[10] As shown in Table 1 below, DRF obligations have increased considerably since 1990 in comparison to those recorded in previous decades.

Table 1. Disaster Relief Fund, FY1974-FY2005 (millions of dollars, 2005 constant dollars)

		Total Appropriations				Outlays	
FY	[a]Request.	Original	Supplemental	Nominal	Constant	Nominal	Constant
1974	100	200	233	433	1,412	250	816
1975	100	150	50	200	591	206	609
1976	187	187	0	187	517	362	999
1977	100	100	200	300	770	294	754
1978	150	115	300	415	997	461	1,108
1979	200	200	194	394	876	277	616
1980	194	194	870	1,064	2,175	574	1,173
1981	375	358	0	358	668	401	746
1982	400	302	0	302	526	115	201
1983	325	130	0	130	217	202	337
1984	0	0	0	0	0	243	391
1985	100	100	0	100	156	192	299
1986	194	100	250	350	533	335	511
1987	100	120	[b]0	120	178	219	325
1988	125	120	0	[c]120	173	187	269
1989	200	100	[d]1,108	1,208	1,674	140	194
1990	270	98	[e]1,150	1,248	1,668	1,333	1,781
1991	270	0	0	0	0	552	711
1992	[f]184	185	4,136	[g]4,321	5,429	902	1,134
1993	292	292	2,000	[h]2,292	2,816	2,276	2,796
1994	[i]1,154	226	[j]4,709	4,935	5,935	3,743	4,502
1995	320	320	[k]3,275	3,595	4,235	2,116	2,492
1996	320	222	[k]3,275	[k]3,497	4,042	2,233	2,581
1997	320	[l]1,320	[l]3,300	4,620	5,248	2,551	2,898
1998	[m]2,708	320	[n]1,600	1,920	2,155	1,998	2,242
1999	[o]2,566	[p]1,214	[q]1,130	2,344	2,597	3,746	4,149
2000	2,780	[r]2,780	0	2,780	3,019	2,628	2,853
2001	2,909	300	[s, t]	[t]5,890	6,249	3,217	3,413
2002	[u]1,369	664	[v]7,008	[v]12,160	12,677	3,947	4,114
2003	1,843	800	[w]1,426	[w]2,199	2,255	8,541	8,761
2004	1,956	1,800	[x]2,275	[x]2,042	[y]2,068	[y]3,044	[y]3,082
2005	2,151	2,042	[x]8,500	10,542	10,542	[y]3,363	[y]3,363
Total	24,240	16,360	48,988	72,099	84,455	50,648	60,224

Sources: U.S. President, annual budget documents; appropriations legislation; U.S. FEMA budget justifications. Constant dollar amounts based on CRS calculations in turn based on GDP (chained) price index in U.S. President (Bush), *Historical Tables, Budget of the United States Government, Fiscal Year 2005* (Washington, 2004), pp. 184-185. Table prepared by Keith Bea, Specialist in American National Government, Government and Finance Division.

a. Data in the request column generally represent the first budget request submitted by the Administration each year and do not include amended or supplemental requests. Notes in this column provide additional detail.

b. In Feb. 1987, a total of $57.5 million was rescinded and transferred from the DRF to the Emergency Food and Shelter Program account (P.L. 100-6). That amount was returned to the fund the same year in supplemental appropriations legislation enacted in July 1987 (P.L. 100-71).
c. P.L. 100-202, the Continuing Appropriations Act for FY1988, appropriated $120 million for disaster relief. According to FEMA, the original appropriation for that fiscal year was $125 million, but $5 million was transferred to the Department of Labor for "low income agriculture workers."
d. Supplemental funds were included in P.L. 101-100, continuing appropriations legislation enacted after Hurricane Hugo struck in Sept. 1989. According to FEMA, this amount was "referred to as a supplemental but was an increase in the original appropriation during a continuing resolution."
e. P.L. 101-130, enacted after the Loma Prieta earthquake, appropriated $1.1 billion in supplemental funding for FY1990. In addition, $50 million was appropriated in P.L. 101-302, dire emergency supplemental appropriations legislation. Table 1 does not reflect a $2.5 million transfer from the President's unanticipated needs fund.
f. FY1992 request does not include the budget amendment of $90 million submitted by the Administration.
g. Appropriations for FY1992 included a $943 million dire emergency supplemental in P.L. 102-229, enacted in fall 1991 after Hurricane Bob; $300 million after the Los Angeles riots and flooding in Chicago (spring 1992) in P.L. 102-302; and $2.893 billion in P.L. 102-368 after Hurricanes Andrew and Iniki, Typhoon Omar, and other disasters.
h. Total for FY1993 includes the $2 billion supplemental approved after the Midwest floods in 1993 (P.L. 103-75).
i. The original FY1994 budget request was $292 million. On July 29, 1993, a supplemental request of $862 million was sent by President Clinton to Congress.
j. Supplemental appropriations for FY1994 enacted after the Northridge earthquake struck Los Angeles (P.L. 103-211).
k. Additional supplemental appropriation approved for Northridge earthquake costs (P.L. 104-19) for FY1995, with the same amount ($3.275 billion) reserved for a contingency fund for FY1996. However, $1 billion of the contingency fund was rescinded in FY1996 omnibus appropriations, P.L. 104-134. In the same legislation, another $7 million was also appropriated to other FEMA accounts for costs associated with the bombing of the Alfred P. Murrah federal building in Oklahoma City.
l. The FY1998 budget appendix (p. 1047) noted a transfer of $104 million from the disaster relief fund in FY1996. In the FY1997 appropriations act (P.L. 104-204), $1 billion that had been rescinded in FY1996 (P.L. 104-134) was restored, and $320 million in new funds were appropriated. Supplemental appropriations of $3.3 billion were approved in P.L. 105-18 after flooding in the Dakotas and Minnesota, and after storms in other states were declared major disasters. The legislation specified, however, that of the total, $2.3 billion was to be available in FY1998 only when FEMA submitted a cost control report to Congress. This requirement was met, and the funding was made available in FY1998.
m. The FY1998 request consisted of a $320 million base amount plus $2.388 billion "to address actual and projected requirements from 1997 and prior year declarations." (*Budget Appendix FY1998*, p. 1047). Does not include $50 million requested for the DRF for mitigation activities.
n. Supplemental appropriations legislation (P.L. 105-174) for FY1998, approved for flooding associated with El Niño and other disasters.
o. The FY1999 request consisted of $307.8 million for the DRF and an additional $2.258 billion in contingency funding to be available when designated as an emergency requirement under the Balanced Budget Act of 1985, as amended.

p. The FY1999 omnibus appropriations act (P.L. 105-277, 112 Stat. 2681-579) included $906 million for costs associated with Hurricane Georges, flooding associated with El Niño, and other disasters.
q. Emergency supplemental appropriations for FY1999 (P.L. 106-31) included $900 million for tornado damages as well as $230 million for unmet needs, subject to allocation directions in the conference report (H.Rept. 106-143).
r. FY2000 appropriations act (P.L. 106-74, 113 Stat. 1085) included disaster relief funding as follows: $300 million in regular appropriations and $2.480 billion designated as emergency spending for costs associated with Hurricane Floyd and other disasters. In addition, the Consolidated Appropriations Act (P.L. 106-113) authorized the Director of FEMA to use up to $215 million in disaster relief funds appropriated in P.L. 106-74 for the purchase of residences flooded by Hurricane Floyd, under specified conditions.
s. Supplemental appropriations legislation (P.L. 106-246) authorized that $77 million from the DRF was to be used for buyout and relocation assistance for victims of Hurricane Floyd. The act also appropriated $500 million in a separate account for claim compensation and administrative costs associated with the Cerro Grande fire that destroyed much of Los Alamos, New Mexico.
t. P.L. 107-38 appropriated $40 billion in response to the terrorist attacks of Sept. 11, 2001. Pursuant to the statute, these funds for FY2001 were allocated by the Office of Management Budget from the Emergency Response Fund (ERF). Of the total appropriated in P.L. 107-38 after the Sept. 11 attacks, $4.4 billion was allocated for FY2001 through P.L. 107-117 (115 Stat. 2338). The total available for obligation for FY2001 ($5.9 billion) taken from FEMA *Justification of Estimates, FY2003*, p. DR-2.
u. Request for FY2002 did not include funding for the Disaster Relief Contingency Fund.
v. Congress appropriated a total of $7.008 billion for FY2002 in P.L. 107-117 and P.L. 107-206 to meet additional needs associated with the terrorist attacks. Total funds available ($12.16 billion) include a transfer from TSA, $1 billion released from the Emergency Contingency Fund, and other sources. See DHS, Emergency Preparedness and Response Directorate, *Justification of Estimates, FY2004*, p. DR-2.
w. Includes $442 million in P.L. 108-69 and $938 million in P.L. 108-83 to meet needs associated with tornadoes, winter storms, the recovery of wreckage of the Space Shuttle *Columbia* and other disasters. Also, funds appropriated in these measures and in the FY2004 appropriations act for DHS (P.L. 108-90) have been used for costs associated with Hurricane Isabel. Total of $2.199 billion available taken from DHS, Emergency Preparedness and Response Directorate, *Justification of Estimates, FY2005*, p. FEMA-18.
x. P.L. 108-106, which primarily addressed reconstruction costs in Iraq and Afghanistan, also contained an appropriation of $500 million for needs arising from disasters in fall 2003, including Hurricane Isabel and the California fires. Section 4002 of the act designates the funds an emergency requirement pursuant to the budget resolution adopted by Congress (H.Con.Res. 95), but the Consolidated Appropriations Act for FY2004 (Section 102(a), Division H, P.L. 108-199) rescinded $225 million of the $500 million appropriated in P.L. 108-106. Total of $2.043 billion taken from: DHS, Emergency Preparedness and Response Directorate, *Justification of Estimates, FY2005*, p. FEMA-18. P.L. 108-303, enacted after Hurricanes Charley and Frances struck Florida, appropriated $2 billion to the DRF and gave discretion to DHS to transfer $300 million to the Small Business Administration for disaster loans. P.L. 108-324, Division B of the Military Construction Appropriations Act for FY2005, appropriated an additional $6.5 billion to the DRF.
y. Outlay data and constant dollar calculations based on estimates.

Issues for the 109th Congress

Controlling Federal Expenditures

The increase in federal expenditures since 1990 has been the subject of some debate. A report issued by the Office of Inspector General (OIG) for FEMA concluded that the increase in federal disaster costs since 1989 "is due to a greater number and magnitude of disasters, expansion of the law and eligibility for assistance, and interpretation of the law and regulations."[11] Some contend that other factors, notably political considerations, contribute to the costs of disaster relief as well. The author of one study reportedly analyzed data from the insurance industry, climatic study organizations, and DHS, and concluded that "electoral motivations ... had a dramatic effect on which states were granted disaster declarations."[12] More specifically, and less dramatically, Tarcey reported in a published summary of his work: "The best predictor of a disaster declaration, bar none, is actual need. The question arises in these marginal cases, when it's unclear whether to give or not."[13]

Another perspective on the issue was presented in a study completed by the Government Accountability Office (GAO) that also considered the effects of politics on disaster declarations. After examining presidential declaration data from the perspective of the party affiliation of governors and members of state congressional delegations, the authors concluded that there "were no indications that party affiliation affected White House major disaster declaration decisions."[14] In light of concerns about funding decisions after the hurricanes, and the rising deficit, Members of the 109th Congress may elect to consider means of controlling costs or establishing alternative funding mechanisms.

Long-Term Recovery Policy

The Stafford Act authorizes a variety of assistance for stricken states through the full range of events that precede and result from catastrophes. Specific provisions, and the phases of emergency management to which they pertain, include

- **Preparedness:** funding is provided to help states and communities develop disaster preparedness plans, improve warning systems, and conduct training and exercise activities;[15]
- **Emergency response:** federal resources may be used to provide equipment, supplies, and personnel to support state and local efforts, to coordinate disaster relief operations, and provide essential assistance to meet emergency needs;[16]
- **Recovery:** repair of damaged buildings and infrastructure, debris removal, temporary housing and limited home repairs, and revenue loss loans;[17] and,
- **Hazard mitigation**: financial assistance before and after a major disaster declaration is issued to reduce future disaster losses.[18]

The Stafford Act does not explicitly authorize the President to provide long-term recovery assistance to communities.[19] However, the federal government has occupied the field to some extent. The Secretary of Commerce is authorized to undertake disaster economic recovery activities.[20] The *National Response Plan*, the document that sets forth agency responsibilities when major disaster declarations are issued, includes a "Long-Term Community Recovery and Mitigation Annex." This annex "provides a framework for federal government support to state, regional local, and tribal governments, nongovernmental organizations (NGOs), and the private sector designed to enable community recovery from the long-term consequences" of catastrophes.[21] Congress may be called upon to consider whether the existing statutory authority of the Secretary of Commerce is adequate, and to consider whether the Secretary of DHS may undertake long-term recovery activities as set out in the *National Response Plan*, or to consider other legislative initiatives

ENDNOTES

[1] The Robert T. Stafford Disaster Relief and Emergency Assistance Act, 42 U.S.C. 5121 et seq. In addition to the assistance authorized by the Stafford Act, a wide range of aid is provided by other federal agencies under general statutory authority.

[2] For a list of major disaster declarations, see U.S. Federal Emergency Management Agency, "Federally Declared Disasters by Calendar Year," at [http://www.fema.gov/library/drcys.shtm], visited Aug. 29, 2005. For information on the assistance provided after four hurricanes struck Florida in 2004, see CRS Report RL32581, *Assistance After Hurricanes and Other Disasters: FY2004 and FY2005 Supplemental Appropriations*, by Keith Bea and Ralph Chite.

[3] For criteria considered in the declaration of a major disaster, see 44 CFR 206.48.

[4] 42 U.S.C. 5122(2).

[5] "The President may exercise any authority vested in him by ... this title with respect to an emergency when he determines that an emergency exists for which the primary responsibility for response rests with the United States because the emergency involves a subject area for which, under the Constitution or laws of the United States, the United States exercises exclusive or preeminent responsibility and authority. In determining whether or not such an emergency exists, the President shall consult the Governor of any affected state, if practicable. The President's determination may be made without regard to subsection (a) of this section." 42 U.S.C. 5191(b).

[6] One indicator used in determining whether a major disaster declaration will be declared for disasters that occur after Oct. 1, 2004, is whether damages exceed $1.14 per capita in each affected state. See U.S. Department of Homeland Security, Federal Emergency Management Agency, "Notice of Adjustment of Statewide Per Capita Impact Indicator," *Federal Register*, vol. 69, Oct. 19, 2004, p. 61515. Similarly, a "countywide" indicator ($2.84 per capita for disasters that occur after October 1, 2004) is used in determining which local jurisdictions will be included in the major disaster declaration. See See U.S. Department of Homeland Security, Federal Emergency Management Agency, "Notice of Adjustment of Statewide Per Capita Impact Indicator," *Federal Register*, vol. 69, Oct. 19, 2004, p. 61505-61506.

[7] For regulations on the request and declaration process, see 44 CFR §§206.35-206.39.

[8] Eligible private non-profit organizations provide essential services and are open to the general public. Essential services include medical and custodial care, education, water, sewer and electrical systems, homeless shelters, and cultural programs such as those offered by museums.

[9] Such accounts are referred to as "no-year" accounts. For background on this and other types of federal budget accounts, see CRS Report 98-410 GOV, *Fact Sheet on the Budget Process*, by Bill Henniff, Jr.

[10] U.S. Congress, Senate Bipartisan Task Force on Funding Disaster Relief, *Federal Disaster Assistance*, S.Doc. 104-4, 104th Cong., 1st sess., (Washington: GPO, 1995). The House convened a task force that issued an unpublished report. Following completion of the task force efforts, some Members introduced a concurrent resolution (H.Con.Res. 39, 104th Congress) seeking a "fundamental overhaul of federal disaster policies." See also U.S. Congress, House Committee on the Budget, Task Force on Budget Process, *Budgetary Treatment of Emergencies*, hearing, 105th Cong., 2nd sess., June 23, 1998 (Washington: GPO, 1998).

[11] U.S. Federal Emergency Management Agency, at [http://www.fema.gov/library/pp2man.shtm], visited Nov. 19, 2004.

[12] For a summary, see Andrew Reeves, "Plucking Votes from Disasters," *Los Angeles Times*, May 12, 2004, p. A19.

[13] Brian Tarcey, "Flooding the Ballot Box: The Politics of Disaster," *Harvard Magazine*, at [http://www.harvard-magazine.com/on-line/030492.html], visited May 21, 2004.

[14] U.S. General Accounting Office, *Disaster Assistance: Timeliness and Other Issues Involving the Major Disaster Declaration Process*, GAO/RCED-89-138, May 25, 1989, pp. 1, 4.

[15] 42 U.S.C. 5131.

[16] 42 U.S.C. 5170a, 5170b. Even before a major disaster declaration is issued, the President is authorized to direct the Secretary of Defense to use departmental resources for emergency work essential for preserving life and property (42 U.S.C. 5170b(c)).

[17] 42 U.S.C. 5172-5186.

[18] 42 U.SC. 5133 (pre-disaster mitigation); 42 U.S.C. 5170c (post-disaster mitigation).

[19] Title V of P.L. 93-288, the Disaster Relief Act of 1974 (88 Stat. 160-163), authorized the President to provide economic recovery assistance "after the period of emergency aid and replacement of essential facilities and services." Congress never funded this authority, and it was repealed in 1998 amendments (see Sec. 102(c) of P.L. 105-393, 112 Stat. 3617).

[20] 42 U.S.C. 3149(c)(2).

[21] U.S. Department of Homeland Security, *National Response Plan* (Washington: 2004), pp. ESF#14-1 through 14-6.

In: Federal Disaster Programs and Hurricane Katrina
Editor: Douglas D. Syzerhans, pp. 39-97
ISBN 1-59454-889-7

Chapter 7

ORGANIZATION AND MISSION OF THE EMERGENCY PREPAREDNESS AND RESPONSE DIRECTORATE: ISSUES AND OPTIONS FOR THE 109TH CONGRESS

Keith Bea

SUMMARY

On July 13, 2005, Department of Homeland Security (DHS) Secretary Michael Chertoff released an assessment of departmental functions known as the second stage review, or 2SR. The recommended changes, planned for implementation on October 1, 2005, include one to dismantle the Emergency Preparedness and Response (EPR) Directorate of DHS, also referred to as the Federal Emergency Management Agency (FEMA). FEMA would remain within DHS but become a smaller entity reporting directly to the Secretary and would be responsible for consequence management.

The Secretary's recommendation is consistent with mission shifts and gaps evident since the establishment of DHS (and EPR) in 2003. Certain functions and tasks of EPR are no longer administered in the directorate. Some authorities originally transferred into EPR have few resources. In short, EPR has administered a portfolio of authorities more limited than authorized by Congress.

By congressional direction and tradition, FEMA's mission comprises four broad areas — preparedness, response, recovery, and hazard mitigation. Secretary Chertoff identifies response and recovery as the "core" operations of FEMA that will be retained in the agency. Preparedness functions would be transferred to a new directorate. No information is available concerning plans for the disposition of hazard mitigation activities.

Congress might elect to evaluate the Administration's 2SR proposal by reviewing whether authorities set out in the Homeland Security Act for EPR (Title V of P.L. 107-296) should remain the focus of one DHS entity or be integrated into other DHS units as proposed by the Secretary. Congress might also broaden the debate by considering the scope and reach of federal authorities that are missions for entities other than DHS.

Options that might be considered include strengthening EPR/FEMA, endorsing the Secretary's proposal, or reassessing the range of homeland security missions and emergency authorities in departments or agencies other than DHS. Congress would have to amend the Homeland Security Act of 2002 (P.L. 107-296) if all aspects of the

Secretary's recommendation were to be implemented. Several bills pending before Congress might serve as legislative vehicles. H.R. 1817, which would authorize appropriations and establish new preparedness authorities, might be considered appropriate legislation for such changes. H.R. 3477 would direct the DHS Secretary to establish regional offices, an important element in coordinating federal and state activities. Perhaps of greatest significance, some Members of Congress are reportedly considering new legislation in the wake of the tragic events at the end of August, 2005, that occurred after Hurricane Katrina. For example, S. 1615, the text of which is not currently available, would establish FEMA as an independent agency.

BACKGROUND

The Administration Proposal

Shortly after his confirmation on February 15, 2005, as Secretary of the Department of Homeland Security (DHS), Michael Chertoff initiated a study, referred to as the Second Stage Review (2SR), of the mission and structure of the department.[1] On July 13, 2005, Secretary Chertoff released his reorganization recommendation; the Administration expects to implement the changes on October 1, 2005.[2]

Months before the release of the findings, Secretary Chertoff reportedly testified before the House Government Reform committee that the review is intended to generate "results without regard to bureaucratic stovepipes...that shares information effectively both up and down the ranks of the department, and externally, with our federal, state, local and private sector partners."[3] Secretary Chertoff's 2SR recommendations reflect elements of a similar proposal presented in a report issued at the end of 2004 by the Heritage Foundation.[4] The report included a recommendation to consolidate "critical infrastructure protection, preparedness, and state/local/private coordination efforts under an Undersecretary for Protection and Preparedness."[5] If implemented, the Secretary's recommendations will lead to significant changes in DHS operations and lines of authority.

Under the Secretary's proposal, the Emergency Preparedness and Response (EPR) Directorate of DHS would be dismantled, with preparedness functions moved to a new Preparedness Directorate. The Federal Emergency Management Agency (FEMA), currently part of and synonymous with EPR, would become a separate DHS entity that would report directly to the Secretary. FEMA would retain responsibility for consequence management after catastrophes occur. In testimony before a House committee the Secretary presented the rationale for realigning FEMA's functions as follows:

> What the restructuring proposes to do is to take out of FEMA a couple of elements that were really not related to its core mission, that were more generally focused on the issue of preparedness in a way that I think was frankly more of a distraction to FEMA than an enhancement to FEMA. Obviously, FEMA's expertise as a response and recovery agency and as an operational agency, is very, very important to our preparedness effort, as is the expertise of a number of our components, like Secret Service or Coast Guard, which are also going to be, obviously, working very closely with our preparedness component.
>
> But we wanted to make sure that FEMA was, as an operational agency, capable of focusing on its core mission, that it was a direct report to the secretary so it gets the direct attention that it needs. And we wanted to make sure the leadership of FEMA was not torn

between its need to focus on the FEMA role and these additional, rather more strategic, preparedness functions, which we think that we are now seeking to unify and put together in a coordinated fashion.[6]

Through the 2SR process, Secretary Chertoff is seeking to build a more unified and focused department. The proposed transfer of preparedness and certain response functions from EPR, and the elimination of the directorate, arguably is one means of achieving that goal. The proposal to eliminate EPR, retain FEMA as a smaller entity with fewer responsibilities, and create two new organizational components — the Office of Operations Coordination (OOC) and the Preparedness Directorate (PD) — would, according to the Secretary, result in a more focused alignment of organizations and missions.

Under the Secretary's proposal, FEMA would report directly to the Secretary and would continue to administer federal response and recovery authorities after catastrophes occur. OOC would "provide the Secretary with improved crisis and operational management tools" and include the Homeland Security Operations Center (HSOC), the staff of which monitors threats and incident management operations. The Preparedness Directorate would exercise administrative responsibility for preparedness and training functions currently held by EPR and other DHS entities.[7] The U.S. Fire Administration, hazardous material training, the chemical stockpile, the radiological emergency preparedness programs, and BioShield would be transferred to PD.

This report provides background information on matters relevant to the proposal to eliminate EPR, shift the remaining preparedness functions to a new directorate, and refocus FEMA's mission solely to consequence management. Because the 2SR initiative makes no mention of the effect of the proposal on FEMA's hazard mitigation efforts, this report suggests that this is an issue that Congress might elect to investigate.

This report provides information on the authorities and missions of EPR and reviews actions taken since the establishment of DHS to modify the directorate's functions. In addition, the report examines the homeland security and emergency management authorities that Congress has assigned to federal entities other than DHS. It presents as comprehensive a picture as possible of relevant authorities and administrative issues. The report concludes with options that Congress might elect to consider as it evaluates the merits of Secretary Chertoff's 2SR proposal.

The Emergency Preparedness and Response Directorate

The Homeland Security Act of 2002 (HSA) established the Emergency Preparedness and Response (EPR) Directorate in DHS.[8] Title V of the HSA transferred the functions, personnel, resources, and authorities of six existing entities into EPR, as shown below:[9]

1. the Federal Emergency Management Agency (FEMA), except terrorism preparedness;[10]
2. the Integrated Hazard Information System (IHIS), previously administered by the National Oceanic and Atmospheric Administration (NOAA) of the Department of Commerce;[11]

3. the National Domestic Preparedness Office (NDPO) of the Federal Bureau of Investigation and related functions of the Attorney General;[12]
4. the Domestic Emergency Support Teams (DEST) of the Department of Justice and related functions of the Attorney General;[13]
5. the Office of Emergency Preparedness (OEP) and related functions of the Secretary of the Department of Health and Human Services (HHS) and the Assistant Secretary for Public Health Emergency Preparedness;[14] and,
6. the Strategic National Stockpile (SNS) of HHS and related functions of the Secretary.[15]

A seventh capability, the Nuclear Incident Response Team (NIRT), is organized, equipped, and trained by the Department of Energy and the Environmental Protection Agency and operates, as directed by the DHS Secretary, as an organizational unit of EPR.[16] In addition to these functions, the statute also sets forth specific responsibilities for EPR that include the following:

- promoting the effectiveness of emergency responders;
- supporting NIRT through standards, training exercises, and the provision of funds;
- managing, overseeing, and coordinating specified federal response resources;
- aiding disaster recovery;
- creating an intergovernmental national incident management system;
- consolidating existing federal response plans into one plan;
- ensuring that emergency responders acquire interoperative communications technology;[17]
- developing a coordinated strategy for public health-related activities; and
- using private sector resources.[18]

The provision of the HSA that appears most pertinent to the Secretary's 2SR proposal is the section titled "Role of the Federal Emergency Management Agency." Section 507 charges FEMA with "carrying out its mission to reduce the loss of life and property and protect the Nation from all hazards by leading and supporting the Nation in a comprehensive, risk-based emergency management program."[19] Whereas much of FEMA's authority initially rested on executive directives that transferred functions and resources, this statutory provision explicitly stated the broad reach of FEMA's mission.

Table 1 presents summary information on the tasks currently administered by major components of EPR under this statutory authority. Figure 1 presents an organization chart of the major EPR components. Figure 2 presents an organization chart of DHS to show EPR within the context of the entire department. The information in the organization charts is based on data available on the DHS website and from other sources.[20]

Table 1. Summary of Components of Emergency Preparedness and Response Directorate

Division or Office	Primary Responsibilities
Administrative Offices	
Under Secretary	Administers EPR and serves as director of FEMA
Policy Office	Develops and monitors implementation of policy and considers need for policy changes
External Affairs Office	Coordinates distribution of information to external entities
Regional Operations Offices	Regional and area offices serve as liaison with state, territorial, and local governments throughout the nation
Plans and Programs Office	Develops and monitors implementation and goals strategies
Program Units	
National Security Coordination Office	Provides leadership to federal agencies for continuity of operations (COOP), develops and implements exercises for the continuity of government (COG) program
U.S. Fire Administration	Provides leadership and support of efforts to prevent and control fires and enhance emergency medical services
Divisions	
Preparedness	Develops national response capability, sponsors tabletop exercises, enhances capabilities to respond to incidents at U.S. Army chemical stockpile sites, helps monitoring efforts around nuclear power plants, assesses capabilities of units of government, maintains and refines the National Incident Management System (NIMS) and components
Mitigation	Works with state and local units of government to reduce the risks of hazards from future disasters, updates flood maps, administers the pre-disaster mitigation grant program
Response	Integrates DHS response teams, deploys Federal Initial Response Support Teams (FIRSTs), improves disaster response and recovery initiatives, develops catastrophic disaster response plans in high-risk communities, improves hospital surge and mass patient care capabilities
Recovery	Leads efforts to rebuild communities after catastrophes, develops and implements plans to expedite aid after catastrophic disasters, improves decontamination efforts, administers debris removal program, coordinates efforts to restore public services

Sources: U.S. Department of Homeland Security, Emergency Preparedness and Response Directorate, Federal Emergency Management Agency, *Fiscal Year 2006 Congressional Justification* (Washington: 2005), pp. FEMA-1through FEMA-7; information based also on conversations between the author and FEMA congressional liaison staff.

Emergency Preparedness Authorities

Title V of the HSA identifies the mission and authority of EPR, but actions taken by the Administration and by Congress since the establishment of DHS have resulted in a mismatch whereby the statutory authority does not match the mission carried out by directorate officials. The following subsection provides background information on the preparedness authorities currently assigned to EPR in the statute, and tracks the transfer of responsibilities from the directorate.

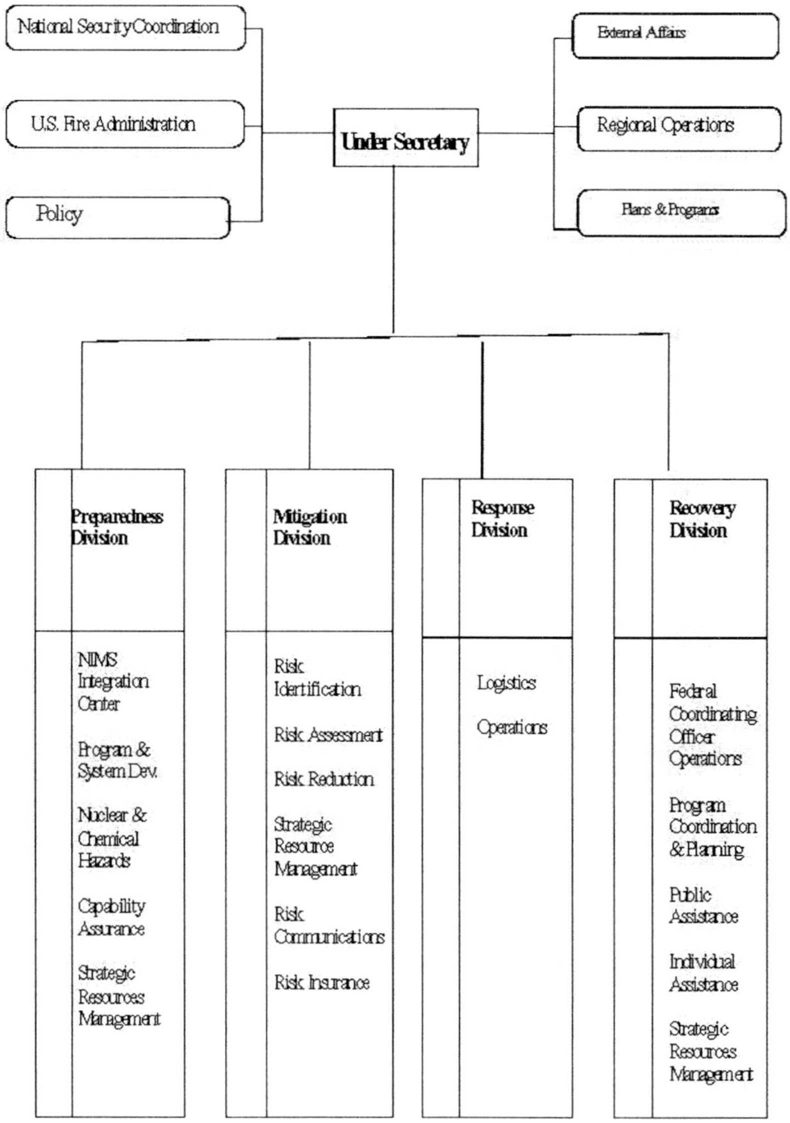

Sources: CRS, based on information provided by the FEMA Office of Legislative Affairs, and organization charts dated June 28, 2004, and Dec. 2, 2003.

Figure 1. Emergency Preparedness and Response Directorate Organization Chart, 2005

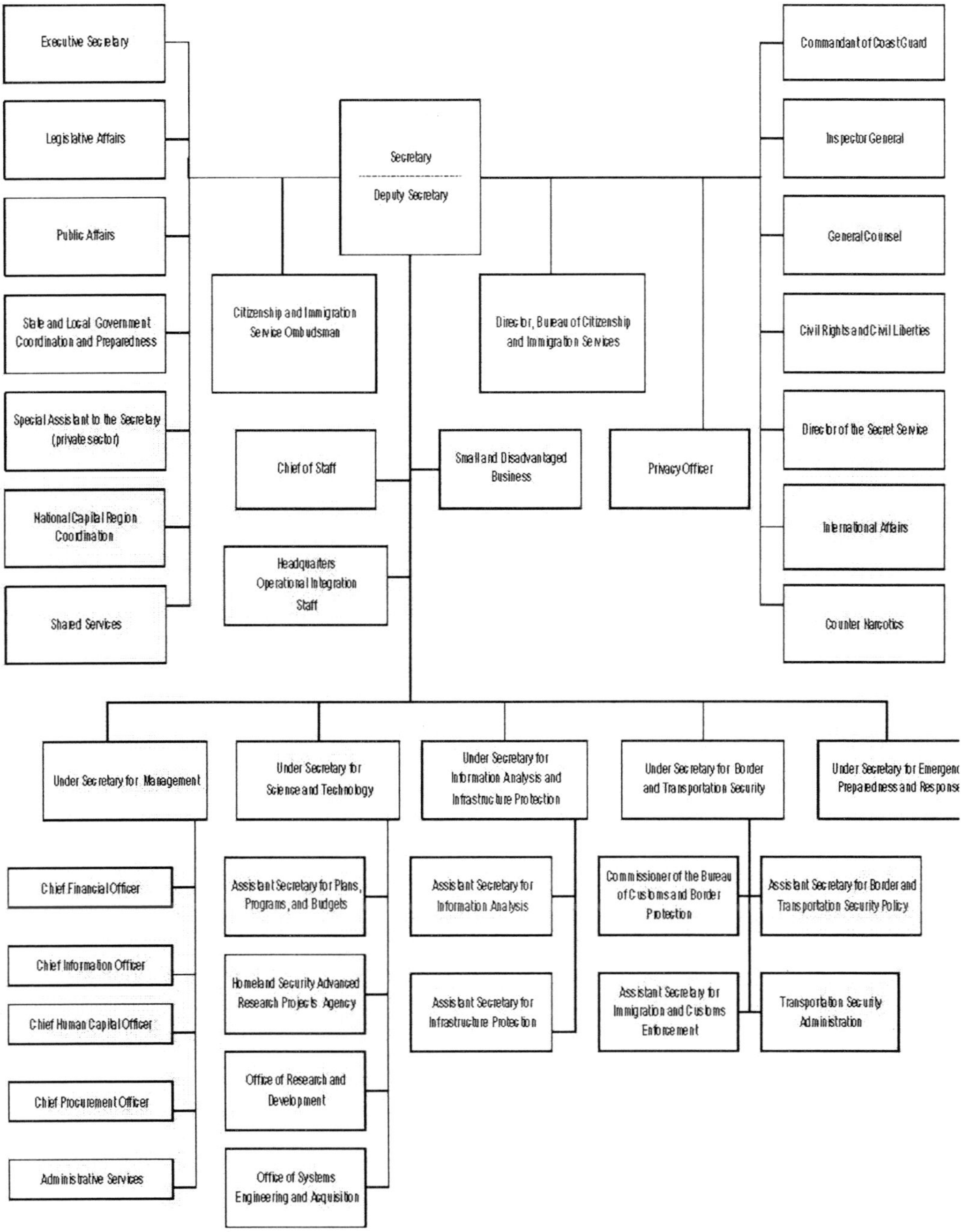

Sources: DHS organization chart prepared by Henry Hogue, Analyst in American National Government, and Mildred Boyle, Research Production Assistant, Government and Finance Division, CRS, April 6, 2005. Chart based on information available from the DHS website [http://www.dhs.gov/interweb/assetlibrary/DHS_OrgChart_2004.pdf] and U.S. National Archives and Records Administration, Office of the Federal Register, *The United States Government Manual 2004-2005* (Washington: GPO, 2004), p. 234. DHS personnel verified the accuracy of information in the chart in April 2005.

Figure 2. Department of Homeland Security Organization Chart

Prior to its incorporation into DHS, FEMA administered preparedness responsibilities through offices and directorates such as the Plans and Preparedness Directorate (generally from the inception of the agency through the early 1980s during the Carter Administration); national preparedness and planning entities during the Reagan Administration (the late 1980s through 1992); the Preparedness, Training, and Exercises Directorate (the mid-1990s during the Clinton Administration); and the Office of National Preparedness (the early years of the George W. Bush Administration). Throughout FEMA's existence, emergency preparedness has been an integral component of the agency's functions.

The Bush Administration sought to maintain a preparedness function, including terrorism preparedness, in FEMA as part of its initial plan for DHS. However, during congressional debate on the HSA, terrorism emergency preparedness missions were separated from FEMA and brought under the jurisdiction of the Office for Domestic Preparedness (ODP) within the Border and Transportation Security Directorate.[21]

This action proved a precursor to subsequent decisions to decrease the emergency preparedness mission of EPR. On March 26, 2004, the Secretary of DHS reorganized the department and transferred ODP to the Office for State and Local Government Coordination and Preparedness (SLGCP) within the Office of the Secretary.[22] As part of this consolidation, and as approved by Congress, responsibility for administering the following programs migrated from EPR to SLGCP:

- Assistance to Firefighters program,
- Emergency Management Performance Grant program,
- first responder counter-terrorism training assistance,
- state and local all-hazards emergency operations planning,
- Citizens Corps,
- interoperable communications equipment,
- Community Emergency Response Teams, and
- Metropolitan Medical Response System (MMRS).[23]

In addition to the transfer of grant authority to SLGCP, Congress and the administration agreed to transfer three other preparedness missions from EPR. First, as noted on page 4 of this report, Congress reversed the HSA provision that brought the Strategic National Stockpile, which includes the preparation of pharmaceuticals, vaccines, and medical supplies ready for deployment, into EPR, and returned the authority for the stockpile to HHS. Second, the 108th Congress removed the FEMA director as coordinator of the National Earthquake Hazards Reduction Program by transferring that authority to the National Institute of Standards and Technology in the Department of Commerce.[24] A third reduction occurred with enactment of the Intelligence Reform and Terrorism Prevention Act of 2004. The HSA originally authorized the Under Secretary of EPR to develop a comprehensive interoperable communications technology program and ensure that first responders obtain the technology. The first element of that function (development of a program) has shifted to ODP as a result of a provision included in the Intelligence Reform Act; the 2004 statute does retain authority in EPR to ensure that first responders acquire the technology.[25]

In summary, since enactment of the HSA, 11 preparedness functions or authorities have been transferred from EPR. The 2SR recommendation to transfer the remaining preparedness authorities out of EPR arguably is not a radical shift in policy and continues recent practices of Congress and the Bush Administration to reduce the preparedness mission of EPR/FEMA. What might be more significant, however, are two other emergency preparedness functions that involve EPR, those involving contingency planning for the continued operation of the government and hazard warning systems.

Continuity of Government Operations

Authorities governing arrangements for the continued operation of the federal government in the event of a national emergency or catastrophe are specified in law, policy, and plans, some of which are not public information given their sensitive and contingent status. These authorities provide for the security and preservation of the senior elected and appointed officials of all three branches of the federal government, and the reconstitution of departments and agencies following an operational interruption. Continuity of government operations plans are designed to ensure the survival of a constitutional form of government and the continuity of essential federal functions.[26]

Another federal preparedness issue related to the mission of EPR concerns the process used to establish federal contingency plans to ensure the continuity of operations (COOP) of federal agencies, and the role that DHS — specifically EPR — exercises in this process.[27] An executive order issued by President Reagan, which serves as a principal authority for federal contingency planning, requires that each federal agency mobilize for, respond to, and recover from a national security emergency.[28] This executive order, as amended, charges EPR with 12 functions related to federal emergency preparedness, including (1) coordinating and supporting federal emergency preparedness programs and plans, (2) coordinating and implementing COOP plans for the federal government, guiding and assisting non-federal planning efforts, and (3) coordinating exercises related to national security.[29] Under this authority, EPR responsibility has been summarized by one Administration official in testimony before Congress as follows:

> FEMA, through my office [Office of National Security Coordination, or ONSC], serves as the lead agent for the federal executive branch's continuity of operations (COOP) and Continuity of Government (COG) programs and as the executive agent for the national-level Emergency Alert System (EAS)....As such, we are working in close cooperation with the Information Analysis and Infrastructure Protection (IAIP) directorate to facilitate coordinated efforts within the department.[30]

In order to carry out this mission, FEMA issued the *Federal Preparedness Circular (FPC) 65* that "provides guidance to federal executive branch departments and agencies for use in developing contingency plans and programs for continuity of operations (COOP)."[31] The circular states the mission of ONSC is to "formulate guidance and establish common standards for agencies to use in developing viable, executable COOP plans; facilitate interagency coordination as appropriate; and oversee and assess the status of COOP capabilities of federal executive branch agencies."[32] Federal agencies must undertake COOP activities to assure the continuation of essential services in the event of a disaster.[33]

The 2SR documentation released by the Administration does not specify whether ONSC would remain in FEMA or transfer to the new Preparedness Directorate (PD) or the Office of Operations Coordination (OOC). It might be argued that the COOP functions could be viewed as extensions of either of the new entities, or they might be subsumed by the

Secretary. Despite the assignment of duties in E.O. 12656 to FEMA, it would be difficult to argue that this preparedness function should remain with the response and recovery missions presented in the 2SR proposal. Congress might wish to obtain further information to evaluate the possible impact of transferring COOP and COG functions out of EPR.

Hazard Warning Systems

Another preparedness function related to the EPR mission that could become part of the debate concerns the role of FEMA or other DHS entities in the development of warning systems. The Homeland Security Act (HSA) contains two authorities pertinent to hazard warning systems. First, Title II authorizes the Under Secretary of Information Analysis and Infrastructure Protection to administer the Homeland Security Advisory System (HSAS) that provides warnings and information to the public and to state and local governments.[34] Second, Title V authorizes the Under Secretary of EPR to administer a different warning system, the Integrated Hazard Information System (IHIS). The IHIS, or FIRESAT system, as it was renamed by the HSA, was intended to enhance the preparedness functions by improving "efforts to identify threats (specifically wildfires) as soon as possible." However, as already noted, IHIS provided no resources to the directorate when it was transferred to EPR.[35]

Throughout the 108th Congress, neither the Administration nor Congress evinced interest in this authority. Considerable interest, by comparison, focused on the shortcomings of the HSAS, the warning system outside EPR's jurisdiction.[36]

Interest in the issue of the efficacy of hazard warning systems increased in the opening days of the 109th Congress in the aftermath of the tsunami that struck 12 countries in southeast Asia and killed approximately 250,000 persons on December 26, 2004. The Senate has acted on legislation to improve warning systems.[37] The documentation on the 2SR recommendation does not address the Administration's plans for FIRESAT, the HSAS, or warning systems generally.

Emergency Response Authorities

In addition to the preparedness matters reviewed above, Members of Congress might elect to consider the impact of the 2SR reorganization on the response functions. Secretary Chertoff has indicated that response and recovery will remain as the functions to be administered by FEMA. Compared to the seemingly constant criticism leveled at FEMA in the 1980s and early 1990s, the disaster response efforts of FEMA were generally praised. After initial difficulties encountered in the early years of the agency's existence, FEMA gained a reputation for being a successful coordinator and provider of response and recovery operations. However, in 2004 questions were raised about aspects of the response to the four hurricanes that struck Florida.[38] These concerns, however, pale in comparison to the questions being raised about the response to Hurricane Katrina in the Gulf States. Secretary Chertoff's pronouncement that response and recovery missions represent the "core" elements of FEMA might arguably reflect a viewpoint that the agency is a primary provider of relief after the President issues a major disaster or emergency declaration and exercises a lesser role in other emergency management fields.

Since enactment of the HSA, some changes have taken place in the EPR response functions, and questions might be raised about others. The following subsections review two issues associated with the response authorities of EPR, the development and implementation of the *National Response Plan* and the vitality of EPR/FEMA response entities.

National Response Plan

The framework that guides the federal response efforts after a catastrophe overwhelms state and local authorities is the *National Response Plan* (NRP).[39] EPR retains primary responsibility for administration of the NRP, an interagency agreement that assigns responsibilities for activities should the President issue a major disaster or emergency declaration under the Robert T. Stafford Disaster Relief and Emergency Assistance Act (the Stafford Act),[40] or determine that an "Incident of National Significance" has occurred.[41] According to Secretary Chertoff's announcement on August 30, 2005, Hurricane Katrina resulted in the first declaration that an Incident of National Significance had occurred.

The NRP was released by Secretary Ridge in December 2004 after a period of consultation with federal and non-federal officials. While the HSA authorized the EPR Under Secretary to assume responsibility for development of the National Response Plan (NRP),[42] reports indicate that other DHS executives reportedly exercised leadership in this matter.[43] Despite the congressional mandate in Title V of the HSA that development of the NRP would be an EPR (FEMA) responsibility, the exact role of directorate officials in the national plan might be subject to question. Members of Congress might elect to investigate the compliance of DHS officials with the statutory requirement concerning the development of the NRP, and whether other DHS officials who exercised leadership in the development of the plan might be tasked under the reorganization with its implementation. For example, one might argue that the proposed Office of Operations Coordination (OOC) could exercise a role in the implementation of the NRP.

Emergency Response Entities

In contrast to the disagreement in 2002 over whether terrorism preparedness activities would remain in FEMA, the 107th Congress and the Administration agreed to keep response activities in FEMA. It might be argued that Congress intended to more fully integrate federal response capabilities in EPR. The statute transferred the functions and assets of the National Domestic Preparedness Office (NDPO), the Domestic Emergency Support Team (DEST), the Office of Emergency Preparedness (OEP), and, to a limited extent, the Nuclear Incident Response Team (NIRT).[44] The consolidation of these response capabilities into EPR contrasts with the decision to remove terrorism preparedness from FEMA's jurisdiction and indicates support for the directorate's response mission.

Questions might be raised, however, about the extent to which these resources provide value to the response mission of EPR. The HSA transferred authority for DEST and NDPO to EPR, but scant information exists on the plans for deploying DEST, and unknown resources are attached to the operations or needs of the teams. The NRP provides that "Nothing in the NRP alters the existing DEST concept of operation or affects the mission of the DEST to support the FBI SAC [special agent in charge] at the scene of a weapons of mass destruction (WMD) threat or incident."[45] The NRP does not discuss a role for the NDPO.

Since the Secretary has asserted that FEMA will retain response authority under the 2SR recommendation, one might surmise that implementation will not affect the DEST role in the

NRP. However, no information has been released by DHS to indicate whether the DEST function will be reinforced or supported by DHS if it remains in FEMA after the reorganization, or whether modification of the HSA will be sought to change the DEST reference in the statute.

Under the *Concept of Operations Plan* (generally referred to as CONPLAN), which has been superseded by the NRP, the FBI was authorized to "form and coordinate the deployment of a DEST with other agencies, when appropriate, and seek appropriate federal support based on the nature of the situation."[46] Under CONPLAN, the interagency aspects of DEST could be used to help the FBI Special Agent-in-Charge (SAC) at the scene of a disaster understand federal capabilities available for defusing terrorist threats, including those involving chemical, biological, or nuclear weapons.[47] In the past, DEST interagency resources were used to enhance crisis management activities, such as information management and technical assistance, and to meet equipment needs.

The HSA vests responsibility for managing and coordinating the federal response (including DEST, the Strategic National Stockpile, NDMS, NIRT, and overseeing the Metropolitan Medical Response System, or MMRS) to disasters and terrorist attacks with the Under Secretary of EPR.[48] Under HSPD-5, the Secretary of DHS "is the principal federal official for domestic incident management."[49] HSPD-5 also provides that the directive does not alter the authority of federal officers to perform their statutory duty, and that the Attorney General "has lead responsibility for criminal investigations" of terrorist threats and actions. Congress might elect to examine whether the DEST or NDPO functions would change or be augmented under the proposed reorganization.

Also, the Secretary has announced that a Domestic Nuclear Detection Office (DNDO) would report directly to the Secretary. The DNDO would "develop and deploy the next generation of systems that will allow us to detect and intercept a nuclear threat."[50] Congress might elect to consider the relationship of the DNDO to the coordination of NIRT resources.

Recovery Authorities

Concerns have been expressed that FEMA's role in the recovery process is limited to the short term and deficient on the long-term needs of communities.[51] Congress might elect to consider the need for legislation to authorize FEMA's long-term recovery efforts.[52] The 2SR documentation provided by the Secretary of DHS does not provide information on how the proposal might affect the recovery mission of FEMA.

Hazard Mitigation

Federal hazard or catastrophe mitigation policies have been enacted by Congress or created through administrative action for decades. Three hazard programs are administered by FEMA — the Hazard Mitigation Grant Program (also referred to as Section 404),[53] the Pre-Disaster Mitigation program,[54] and the Flood Mitigation Assistance program.[55] The documentation released by Secretary Chertoff on the 2SR initiatives provides no information on the disposition of these, or other, hazard mitigation activities administered within EPR.

Authorities of Other Federal Agencies

As noted in the discussion on the evolution of emergency management policy later in this report, neither the consolidation of authorities into FEMA in 1978 nor the consolidation into DHS in 2003 brought a "comprehensive emergency management" and "all hazards" policy framework into one administrative entity. While some authorities were consolidated into FEMA, many remained vested in operational agencies. One means of identifying the homeland security authorities of agencies other than DHS is the list of "Sector-Specific Agencies" identified by presidential directive as retaining the following policy responsibilities:

- Department of Agriculture — agriculture, food (meat, poultry, egg products);
- Department of Health and Human Services — public health, healthcare, and food (other than meat, poultry, egg products);
- Department of Energy — energy, including the production, refining, storage, and distribution of oil and gas, and electric power except for commercial nuclear power facilities;
- Department of the Treasury — banking and finance;
- Department of the Interior — national monuments and icons; and
- Department of Defense — defense industrial base.[56]

This list of sector specific agencies is not comprehensive. Homeland security authorities arguably related to the mission of DHS (and arguably extensions of the normal operations of other departments or agencies) remain dispersed among the federal agencies. Table 2 identifies statutory authorities that endow entities other than DHS with emergency preparedness responsibilities. The statutory authorities listed in Table 2 provide a broader picture of the homeland security or emergency management authorities exercised by federal entities other than DHS.[57]

Preparedness Authorities in Other Agencies

Counter-terrorism training programs and activities are administered by six departments other than DHS.[58] At least two departments have been reorganized to include emergency management functions at the secretarial level to manage and lead emergency management policies and authorities — the Office of Intelligence, Security, and Emergency Response within the Department of Transportation, charged with administering the emergency preparedness and response duties for the department,[59] and the Office for Public Health Emergency Preparedness within the Department of Health and Human Services (HHS), which "directs and coordinates HHS-wide" efforts related to bioterrorism and public health emergencies.[60] According to HHS officials, President Bush has designated the latter as "the principal federal agency for planning and coordinating response to mass casualty incidents."[61] In addition to these offices, the statutory authorities of agencies other than EPR include preparedness functions related to their basic mission.

Table 2. Statutory Emergency Authorities of Federal Agencies Other Than Department of Homeland Security

Organization or Official	Citation	Task or Authority
Dept. of Agriculture	7 U.S.C. 1926a	emergency water infrastructure aid
	7 U.S.C. 1961	disaster loan
	7 U.S.C. 2273	search and rescue assistance
	16 U.S.C. 2106	fire suppression
	16 U.S.C. 2201	repair from winds
Dept. of Commerce	16 U.S.C. 1455	coastal flood management
		economic assistance
	42 U.S.C. 3149	disaster recovery assistance
	42 U.S.C. 3192	recovery information
National Institute of Standards and Technology	15 U.S.C. 7301	building standards
National Oceanic and Atmospheric Admin.	15 U.S.C. 313c	flood warning
Dept. of Defense	10 U.S.C. 138	homeland security coordination
	10 U.S.C. 371	law enforcement assistance
	10 U.S.C. 382	weapons of mass destruction
	32 U.S.C. 503	seismic vulnerability
	50 U.S.C. 2301	emergency preparedness assistance
	50 U.S.C. 2314	emergency response team
Corps of Engineers	33 U.S.C. 426p	flood emergency aid
	33 U.S.C. 467	dam safety
	33 U.S.C. 701n	emergency response
	33 U.S.C. 709a	flood hazards
	33 U.S.C. 2332	flood hazards
	33 U.S.C. 2240	port emergencies
	33 U.S.C. 2293	civil works management
Dept. of Education	20 U.S.C. 1065	emergency fund use
	20 U.S.C. 6337	emergency waiver authority
	20 U.S.C. 7138	school crises
	20 U.S.C. 7217	emergency waiver authority
	20 U.S.C. 7428	emergency waiver authority
	20 U.S.C. 9251	emergency waiver authority

Table 2. Continued

Organization or Official	Citation	Task or Authority
Dept. of Energy	16 U.S.C. 824a(c)	energy emergencies
	42 U.S.C. 6323	energy emergencies
	42 U.S.C. 7270c	facility vulnerability
	42 U.S.C. 7274d	emergency training
	42 U.S.C. 10137	emergency training
Dept. of Health and Human Services	42 U.S.C. 247d	public health emergency
	42 U.S.C. 243	quarantines, public health plans
	42 U.S.C. 239	smallpox response
	42 U.S.C. 249	medical care for those quarantined
	42 U.S.C. 267	quarantine stations
	42 U.S.C. 300hh	national stockpile
	42 U.S.C. 8621	emergency energy aid
	42 U.S.C. 1320b	waiver authority
	42 U.S.C. 3030	elderly assistance
Dept. of Housing and Urban Development	12 U.S.C. 1701n	reduce attack vulnerability
	12 U.S.C. 1709	mortgage assistance
	42 U.S.C. 3539	disaster fund
	42 U.S.C. 5306	reallocation of funds
	42 U.S.C. 5321	waiver authority
	42 U.S.C. 12750	matching fund waiver
Dept. of the Interior	16 U.S.C. 1011	watershed protection
	42 U.S.C. 5204	disaster recovery plans
	43 U.S.C. 502-503	emergency fund for reclamation
Public Lands Corps	16 U.S.C. 1723	disaster prevention and relief
Dept. of Justice (Attorney General)	20 U.S.C. 7138	school safety
	42 U.S.C. 10501	law enforcement aid
Dept. of Labor	29 U.S.C. 2918	emergency grants
Dept. of Transportation	23 U.S.C. 125	emergency funds
	23 U.S.C. 310	civil defense
	23 U.S.C. 502	seismic vulnerability
	33 U.S.C. 1225	structure protection
	33 U.S.C. 1226	vessel protection
	49 U.S.C. 60132(c)	emergency pipeline response
	49 U.S.C. 5102	hazardous material transportation
	50 U.S.C. 191	vessels in emergency situations

Table 2. Continued

Organization or Official	Citation	Task or Authority
Dept. of the Treasury	19 U.S.C. 1318	emergency authority
	26 U.S.C. 5708	disaster loss aid
	29 U.S.C. 1148	waiver authority
	29 U.S.C. 1302	waiver authority
	42 U.S.C. 2414	flood insurance funding
Dept. of Veterans Affairs	38 U.S.C. 1785	medical assistance
	38 U.S.C. 8117	public health emergencies
	38 U.S.C. 7325	medical response plans
	38 U.S.C. 7326	emergency training
	38 U.S.C. 8105	facility safety
	38 U.S.C. 8111A	health care provision
Corporation for National and Community Service	24 U.S.C. 12576	disaster relief
Environmental Protection Agency	42 U.S.C. 300g	water safety after disasters
	42 U.S.C. 300i	vulnerability assessment
	42 U.S.C. 300j	preparedness grants
	42 U.S.C. 7274d	training grants
	42 U.S.C. 9601	environmental response
	42 U.S.C. 9662	water pollutants and emergencies
	42 U.S.C. 11001	hazardous material releases
Executive Office of the President		
President	42 U.S.C. 217	use of Public Health Service
	42 U.S.C. 5170	declaration authority
	42 U.S.C. 5187	fire suppression
	42 U.S.C. 960	hazardous substance releases
	47 U.S.C. 606(c)	control of radio stations
	50 U.S.C. 2301	weapons of mass destruction
	50 U.S.C. 1621 - 1622	national emergencies
Homeland Security Council	6 U.S.C.A. 491-496	consultation, coordination
National Security Council	50 U.S.C. 2352 - 2353	crisis management

Table 2. Continued

Organization or Official	Citation	Task or Authority
Office of Science and Technology Policy	42 U.S.C. 6613, 6617	advice, consultation
National Aeronautics and Space Admin.	42 U.S.C. 2487	technology for health needs
National Foundation on the Arts and the Humanities		
Institute of Museum and Library Services	20 U.S.C. 9133	waiver authority
National Nuclear Security Administration	50 U.S.C. 2401 - 2402	facility management
Nuclear Regulatory Commission	42 U.S.C. 2242(a)	facility licenses
Office of Personnel Management	5 U.S.C. 3110	employment waivers
Small Business Administration	15 U.S.C. 631(e,g), 636d	disaster loans
U.S. House of Representatives		
Office of Emergency Planning, Preparedness, and Operations	2 U.S.C. 130i	emergency management authority
All departments and agencies		
Agency heads	42 U.S.C. 1856b	emergency fire assistance authorized

Source: CRS examination of federal statutory authority.

Note: Table 2 does not identify presidential directives that assign responsibilities for and establish federal policies pertinent to the mission of EPR. Some of these directives include Executive Orders 12241 (radiological emergencies), 12580 (hazardous substance releases), 12656 (federal emergency preparedness, discussed in this report), 12777 (oil discharges), and 13016 (Superfund amendments).

- The Pipeline and Hazardous Materials Safety Administration of the Department of Transportation administers regulations that direct owners and operators of gas and hazardous liquid pipelines to develop emergency response plans and procedures.[62] The regulations require that responses to accidents and explosions must be coordinated with local public officials and area utilities, to a degree comparable to the coordination requirements established by DHS in the national preparedness system.[63]
- Most recently, Congress and President Bush agreed to enhance the authority of the Department of Transportation to administer activities associated with the transportation of hazardous material. Under Title VII of H.R. 3, the Transportation Equity Act: A Legacy for Users (known as TEA-LU; P.L. 109-59), the Secretary of Transportation administers the Hazardous Materials Emergency Preparedness Fund and is to build a hazardous material incident response system to help first-responders prepare for such incidents.[64]
- The Office of Safe and Drug-Free Schools, U.S. Department of Education, administers the Emergency Response and Crisis Management Grant Program to help schools address "the four phases of crisis planning: Prevention/Mitigation, Preparedness, Response, and Recovery."[65] Local educational agencies that develop plans under this authority must coordinate the results with the appropriate state homeland security plans.

Response Authorities in Other Agencies

In addition to the response authorities that have been, or might yet be, transferred from EPR or led by officials outside of EPR, a wide range of federal authorities related to the mission of EPR are administered by departments and agencies other than DHS. Two categories of authorities might be examined by Congress in light of the 2SR reorganization proposal — the allocation of responsibility for emergency medical policy and the role of military forces in responding to incidents of national significance.

Federal Response Capabilities

A wide range of response teams are operated by federal agencies to expedite assistance after a disaster or attack.[66] The response to a catastrophic event, or an "incident of national significance," as the term is used in the *National Response Plan*, involves a complex series of simultaneous or sequential events involving multiple agencies, levels of government, and non-governmental entities. Many federal agencies, including DHS, are involved in the coordination of federal response teams.

According to the General Accounting Office (now the Government Accountability Office), by late 2000, eight federal agencies hosted 24 teams charged with responding to incidents involving weapons of mass destruction.[67] The establishment of an Office of Coordination (OC) and retention of FEMA with core disaster response authorities may raise questions about the range of federal emergency response capabilities. Such capabilities involve the response teams administered by federal agencies as well as the coordination units charged with ensuring that federal response efforts are used effectively. Table 3 provides summary information on 36 federal response teams and coordination units involved in consequence management. The list is not intended to be comprehensive; it does include the units identified from a search of the sources noted at the end of the table.[68]

Table 3. Federal Consequence Management Response Resources

Team Designation	Summary of Task Responsibilities	Citation for Further Information
Response Teams		
Department of Defense		
Joint Task Force for Civil Support	Plans and integrates DOD support for domestic Chemical, Biological, Radiological, Nuclear, and high yield Explosive (CBRNE) consequence management operations.	[http://www.jtfcs.northcom.mil/pages/mission.html]
Army Special Medical Augmentation Response Team	Provides technical advice concerning hazardous material used in a terrorist event	Part of the U.S. Army Medical Department; see[http://www.army.mil/usapa/epubs/pdf/r40_13.pdf]
Marine Corps Chemical-Biological Incident Response Force	Helps state and local agencies identify WMDs and manage consequences	[http://www.cbirf.usmc.mil/]
Army Radiological Advisory Medical Team	Provides radiological health hazard guidance	Part of the U.S. Army Medical Department; see [http://www.army.mil/usapa/epubs/pdf/r40_13.pdf]
Weapons of Mass Destruction-Civil Support Teams (WMD-CSTs)	Associated with the Army and Air National Guard for each state, assist state and local responders with medical and technical guidance.	[http://www.globalsecurity.org/military/agency/army/wmd-c st.htm]
Air Force Installation Disaster Response Force	Provides Installation Full Spectrum Threat Response support in areas stricken by disasters	[http://www.e-publishing.af.mil/pubfiles/af/10/afi10-2501/afi10-2501.pdf]
Consequence Management Advisory Team (CMAT)	Provides scientific and technical analysis of an emergency	[http://www.dtra.mil/press_resources/fact_sheets/print/index.cfm?factsheet=cm.cf m]
Department of Energy		
Accident Response Group (ARG)	Deploys to the scene of a nuclear weapons accident	[http://www.doeal.gov/opa/Emergency%20Public%20Information/AccResfactsheet.pdf]

Table 3. Federal Consequence Management Response Resources (Continued)

Team Designation	Summary of Task Responsibilities	Citation for Further Information
Response Teams		
Department of Health and Human Services		
National Medical Response Teams include the following need (specific teams)	Decontaminates casualties from a hazardous materialsncident	[http://www.ndms.dhhs.gov/]
-- Disaster Medical Assistance Team (DMAT)	Provides medical care during a disaster	[http://www.ndms.dhhs.gov/dmat.html
-- Disaster Mortuary Operational Response Team (DMORT)	Identifies victims and provides mortuary services after a disaster	[http://www.ndms.dhhs.gov/dmort.html]
-- Veterinary Medical Assistance Team (VMAT)	Provides veterinary services after a disaster	[http://www.ndms.dhhs.gov/vmat.html]
-- National Nurse Response Team (NNRT)	Provides a supply of nurses in response to a WMD event	[http://www.ndms.dhhs.gov/nnrt.html]
Strategic National Stockpile	Transports essential pharmaceuticals in response to public health emergencies	[http://www.bt.cdc.gov/stockpile/]
Chemical/Biological Rapid Response Team	Provides medical care to chemical or biological response units	[http://mmrs.fema.gov/PublicDocs/C-BHMPlan.pdf #search='Chem/Bio%20Rapid%20Deployment%20Team%20OR%20 CBRDT']
Department of Energy		
Nuclear Emergency Support Team (NEST)	Identifies nuclear material, assesses threat, renders material safe, and transports it from site	[http://www.nv.doe.gov/nationalsecurity/homelandsecurity/nest.htm]
Radiological Assistance Program Teams	Evaluates and assists in events that involve radiation risks and hazards	[http://www.lm.doe.gov/rap/program_information.htm]
Radiation Emergency Assistance Center/Training Site (REAC/TS)	Provides direct or indirect medical care to radiation victims	[http://www.orau.gov/reacts/intro.htm]

Table 3. Federal Consequence Management Response Resources (Continued)

Team Designation	Summary of Task Responsibilities	Citation for Further Information
Response Teams		
Department of Homeland Security		
Emergency Response Team (ERT)	Provides for the coordination of federal response and recovery activities, includes ERT-A (advance element to identify state needs immediately after a disaster), Federal Incident Response Support Team (FIRST, a forward component of the ERT-A), and ERT-N (National ERT that deploys for large-scale incidents), and other components that provide specific skills and resources	*National Response Plan*, p. 40
Incident Management Teams(IMT)	Function at state and regional levels as part of an incident command structure at complex disasters	[http://www.usfa.fema.gov/subjects/incident/imt/imt-summary.shtm]
International Medical Surgical Response Team (IMSuRT)	Provides medical and surgical treatment facilities worldwide	[http://www.fema.gov/preparedness/resources/health_med/imsurt.htm]
Urban Search and Rescue Task Forces (US&R)	Provides search and rescue service at the site of a disaster	[http://www.fema.gov/usr/]
U.S. Coast Guard	Provides response and recovery assistance after oil spills on navigable waters	[http://www.uscg.mil/hq/g-m/nmc/response/index.htm#OSRO]
Department of Veterans Affairs		
Medical Emergency Radiological Response Team (MERRT)	Deploys patient treatment and technical advice at the scene of a radiological incident or disaster	[http://www1.va.gov/vhapublications/ViewPublication.asp?pub_ID=310]
Environmental Protection Agency		
EPA Radiological Emergency Response Team (EPA/RERT)	Provides scientific and technical assistance in radiological emergencies	[http://www.epa.gov/radiation/rert/respond.htm]
EPA Environmental Response Team (EPA/ERT)	Provides technical and logistical assistance in response to environmental emergencies and hazardous waste sites	[http://www.ert.org/]
Nuclear Regulatory Commission		
Nuclear Incident Response Teams(NIRT)	Provides personnel and equipment to accompany the federal response to nuclear emergencies	[http://www.dhs.gov/dhspublic/display?theme=17&content=368]

Table 3. Federal Consequence Management Response Resources (Continued)

Team Designation	Summary of Task Responsibilities	Citation for Further Information
Coordination Teams and Capabilities		
Department of Defense		
Air Force Contract Augmentation Program (AFCAP)	Expedites support for military operations other than war (MOOTW) and contingencies arising from disasters or attacks	[http://www.afcap.com/capabilities/capabilities01.htm]l
Department of Energy		
Federal Radiological Monitoring and Assessment Center	In early phase of a radiological emergency, monitors and coordinates assessment of hazard	[http://www.nv.doe.gov/nationalsecurity/homelandsecurity/frmac.htm]
Department of Health and Human Services		
Management Support Teams (MSTs)	Provides support and control for NDMS teams deployed at a disaster site	[http://www.fema.gov/preparedness/resources/health_med/ndms_mst.htm]
Department of Homeland Security		
Forward coordinating team	Responsible for the initial coordination of federal resources at a disaster site	no source identified
Interagency Incident Management Group (IIMG)	Facilitates strategic federal domestic incident management, activated by the Secretary	*National Response Plan,* page 22-24
Joint Field Office (JFO)	Established in the area in which an incident of national significance has occurred; serves as a multiagency center to coordinate response operations; headed by the principal federal official(PFO); and includes state coordinating officer, FBI special agent in charge, a federal coordinating officer (FCO), and sections as needed	*National Response Plan,* page 28-38
Department of Justice		
Strategic Information and Operations Center (SIOC)	Crisis management and operational center operated by at FBI headquarters on a continual basis	[http://www.fbi.gov/hq/siocfs.htm]
Joint Operations Center (JOC)	Operated by FBI at the scene of a crisis to coordinate functional groups deemed necessary with state and local agencies, which could include a DEST representative	

Table 3. Federal Consequence Management Response Resources (Continued)

Team Designation	Summary of Task Responsibilities	Citation for Further Information
Coordination Teams and Capabilities		
Environmental Protection Agency		
On-Scene Coordinators	Monitors and directs responses to oil spills and hazardous material releases	[http://www.epa.gov/superfund/programs/er/nrs/nrsosc.htm]

Sources: CRS examination of various sources, including U.S. Government Accountability Office, Combating Terrorism: Federal Response Teams Provide Varied Capabilities; Opportunities Remain to Improve Coordination, GAO Report GAO-01-14 (Washington: Nov. 30, 2000). U.S. Department of Homeland Security, Federal Emergency Management Agency, National Mutual Aid and Resources Management Initiative Glossary of Terms and Definitions (Washington: July 2005), available at [http://www.fema.gov/pdf/preparedness/507_Mutual_Aid_Glossary.pdf], visited Aug. 18, 2005. Al Mozingo, "Tapping Federal Resources: Activation & Deployment of Federal Assets to WMD Incidents," Homeland First Response, vol. 2, Sept./Oct. 2004, pp. 28-31. Amy E. Smithson and Leslie-Anne Levy, *Ataxia: The Chemical and Biological Terrorism Threat and the U.S. Response* (Washington: The Henry L. Stimson Center, Oct. 2000). U.S. Federal Emergency Management Agency, *Emergency Teams in Federal Disaster Operations*, 9350.1-HB (Washington: July 1999). Telephone conversation with officials associated with the Air Force Civil Engineer Support Agency, Aug. 23, 2005.

In light of the Secretary's proposal to focus FEMA on the response (and recovery) mission, two areas require further discussion — emergency medical response and Department of Defense resources and authorities.

Emergency Medical Response

EPR exercises responsibility for federal emergency health authorities, including serving as the lead federal agency for the Federal Interagency Committee on Emergency Medical Services (FICEMS) and administering the National Disaster Medical System (NDMS).[69] Some related authorities, however, have been retained by the Department of Health and Human Services (HHS). For example, under authority of the Public Health Service Act, the HHS Secretary may take specified action in the event that a "disease or disorder presents a public health emergency."[70] Much of this authority is administered by the Centers for Disease Control and Prevention (CDC). As noted on the CDC website, the activities pursued under the public health emergency statute not only address immediate threats to health (for example, epidemics and chemical or biological attacks), but also include the impacts of explosions and natural disasters.[71] Just as EPR (like FEMA traditionally) has had an "all-hazards" approach to emergencies, the PHSA authority is used to marshal HHS resources for many, if not all, hazards that present significant health threats.

Incorporation of the National Disaster Medical System (NDMS), previously administered by OEP when it was part of HHS, has enhanced EPR's response capabilities. NDMS is a partnership of four departments (DHS, HHS, Defense, and Veterans Affairs) as well as non-federal entities. The transfer of NDMS from HHS, however, may not have resulted in the transfer of comprehensive medical response authority to DHS. For example, HHS announced in early 2005 that its Office of Public Health Emergency Preparedness "is the principal federal agency for planning and coordinating response to mass casualty incidents." The office includes an Office of Mass Casualty Planning responsible for mobile medical units and other resources that appear to be similar to NDMS capabilities as well as the Metropolitan Medical Response System (MMRS) teams funded through the SLGCP office within DHS.[72]

Military Response Activities

Some emergency response duties are vested in the Department of Defense (DOD), including the active duty forces, Armed Forces Reserves, and the National Guard units.[73] The active duty, and reserve forces when activated, respond to enemy threats to defend the United States and are generally authorized to meet "homeland defense" responsibilities set out in Title 10 of the *United States Code.*[74] Army and Air National Guard units serve under authority of their governors, and their state laws, unless called into federal service by the President.[75] Under their state authority, National Guard units have historically helped provide disaster assistance, maintain civil order, and meeting emergency needs.[76] Recent amendments to the federal statute governing the National Guard authorize the Secretary of Defense to issue regulations pertinent to, and reimburse Guard members for actions related to, "homeland defense activity."[77] The conditions under which DOD forces provide support for purposes other than law enforcement are set out in two department directives and other planning documents discussed below.[78]

The Army Reserve and National Guard exercise the following emergency management responsibilities, as summarized by one officer,

> The Army Reserve Component and the National Guard have historically performed missions related to the management of consequences after disasters strike. The Army Reserve provides a wide range of response capabilities in the event of natural or man-made disasters and attacks on the homeland, including almost 200 Emergency Preparedness Liaison Officers to local communities. The Army Reserve has also fielded and trained 75 Chemical Decontamination Platoons with over 2,400 soldiers for mass casualty operations and more than 250 fully equipped Hazardous Material (HAZMAT) technicians to train with local first-responder agencies (i.e. policy, firefighters, HAZMAT teams, emergency medical services, hospitals.[79]
>
> As evidenced by the three devastating hurricanes that hit Florida or the wildfires that blazed through our western states during 2004, the National Guard is a crucial element in a governor's response to natural disasters. Similarly, the National Guard has a prominent role in supporting local and state authorities in their efforts to manage the consequences of a domestic terrorist attack.[80]

Certain law enforcement functions that might be associated with emergency situations (seizure of persons, entering private structures, directing civilian movements) cannot be undertaken by the active duty and reserve armed forces.[81] The lines of authority and responsibility, however, are not always evident. DOD provides considerable support to civil (state and local) authorities overwhelmed by catastrophes.[82] According to the joint force doctrine released by DOD, the department provides military assistance to civil authorities (MACA) to meet consequence management needs in the following three "mission subsets:"

- military support to civil authorities consisting of DOD support for "high-profile emergencies" involving natural disasters, special events, or accidental or intentional manmade disasters;"
- military support to civilian law enforcement agencies that "may include, but is not limited to," national special security events, support for combating terrorism, support to assist in counterdrug operations, maritime security, equipment or facility loans, and intelligence, surveillance, and reconnaissance capabilities;
- military assistance for civil disturbances to suppress insurrections, rebellions, and riots, and help states maintain law and order.[83]

However, the boundaries between the military homeland defense and civil support roles overlaps the homeland security roles of other entities.[84] The overlap has been represented by DOD in the following graphic.

One year after the terrorist attacks of 2001, the Secretary of Defense established an operations component to facilitate the response of the armed services in the United States. The Northern Command, or NORTHCOM, is charged with deterring and defeating attacks in the United States and its possessions. In addition, pursuant to congressional directive, an Assistant Secretary of Defense for Homeland Security has been designated the lead in Department of Defense (DOD) homeland security and homeland defense missions. DOD resources would be available to help DHS respond to the needs of overwhelmed state and local governments.[85]

While federal military resources have proven to be essential elements of response efforts after a catastrophic disaster, questions have been raised about the full extent to which these resources are coordinated with others. As noted by one witness before a Senate committee:

> There is an appropriate role here for the Department of Defense. They've always said that they'll do that. The last thing that I want to see is the Department of Defense figure out how they're going to do catastrophic terrorism on the day after the catastrophic terrorist attack. We need structures and forces in places now that are designed to do this and do this well. I have argued in other places that if you built that kind of capability right in the National Guard that you would actually have a very useful force that could be useful for a range of homeland security missions and would also be very useful for post-conflict operations overseas and would also be used for theater support operations overseas.[86]

Also, as documentation, plans, and strategies for the role of NORTHCOM develop, questions may be raised about the involvement of active duty forces within the United States, and the manner in which those resources are coordinated with DHS units, including FEMA.[87] As noted in the following excerpts from a news report, some DOD officials envision a newer, broader role in domestic emergency situations.

> Several people on the staff here [NORTHCOM headquarters in Colorado Springs] and at the Pentagon said in interviews that the debate and analysis within the U.S. government regarding the extent of the homeland threat and the resources necessary to guard against it remain far from resolved.... William M. Arkin, a defense specialist who has reported on the NORTHCOM's war planning, said the evolution of the Pentagon's thinking reflects the recognition of an obvious gap in civilian resources.[88]

The operational document released by the Joint Chiefs of Staff in August 2005, describes the homeland security framework, mission areas, missions and related supporting operations, and enabling activities" for joint DOD operations.[89]

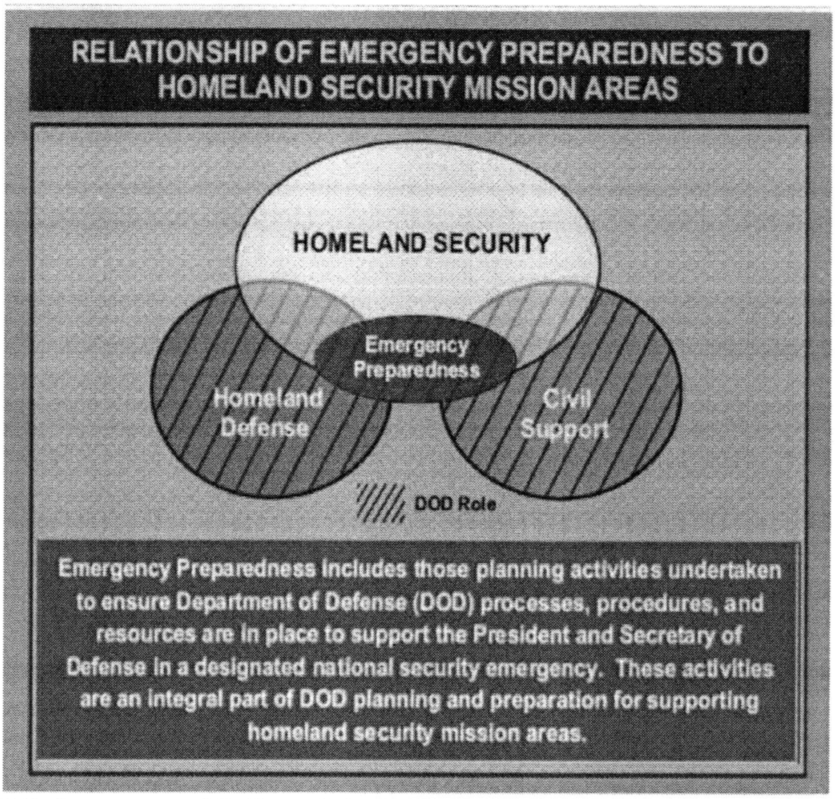

Source: U.S. Department of Defense, Joint Publication 3-26 Homeland Security (Washington: 2005), available at [http://www.fas.org/irp/doddir/dod/jp3_26.pdf], p. I-4, visited Aug. 23, 2005.

Figure 3. DOD Representation of Overlays Among Emergency Preparedness Functions

HSPD-5 states that the directive does not impede the authority of the Secretary of Defense who provides support to civil authorities at the direction of the President. The directive also requires that "appropriate relationships and mechanisms for cooperation and coordination" be developed between DHS and both DOJ and DOD. Additional work may be required to establish fully coordinated relationships between DHS and DOD. As noted in the Joint Armed Forces doctrine released by the Department of Defense:

> Defense support of civil authorities is a new term and is approved in the National Response Plan and is not yet approved in Department of Defense (DOD) policy. Once it is approved in DOD policy, it will likely replace civil support as the overarching term for assistance to civil authorities. However, in the context of this publication, civil support is still used as the overarching term.[90]

Issue Discussion, Questions and Responses

The final report issued by the former Inspector General for DHS (Clark Kent Ervin) at the end of the 108th Congress included the following assessment: "Integrating its many separate components into a single, effective, efficient, and economical department remains one of DHS' biggest challenges."[91] Comments reportedly made by other former, as well as current, DHS officials, Members of Congress, and analysts indicate general agreement on the need to meet this challenge.

Overview of Preparedness Issues

Some would agree with Secretary Chertoff's recommendation that preparedness and emergency operations functions should be taken from EPR and consolidated into the proposed Preparedness Directorate (PD). A new Preparedness Directorate would ensure that planning, training, simulations, and funding are administered together. As summarized by the Secretary, "we believe that preparedness and responder-training functions should be integrated into a dedicated organization."[92] At a hearing held in the opening days of the 109th Congress, an author of the Heritage report that reportedly stimulated much of the 2SR process spoke to the separation of homeland security preparedness activities from response, as follows.

> I would argue for a clear division of responsibilities between operators and supporters.... Preparedness and response, I think, is one of them. Response is clearly an operational function. You want the guy who's in charge of response to be ready to respond, to be thinking about responding and have that be the sole focus of the organization commission. Preparedness, on the other hand, you could argue, is a support function."[93]

One might argue that an important aspect of the proposal is that preparedness functions would continue to be vested in one department, pursuant to the mandate of Congress set out in the HSA and in presidential directives issued by President Bush since September 2001. The proposed transfer of preparedness functions from EPR is consistent with actions taken by Congress and the Administration since the establishment of DHS to divest EPR of preparedness responsibilities and consolidate those authorities in one federal department. Rather than visualizing FEMA as a separate entity that should administer all facets of comprehensive emergency management (as presently called for in the HSA), advocates of the proposal could argue that the consolidation of functions within DHS represents the continued integration of emergency management functions, albeit at a higher administrative level.

Others, however, would disagree with the Secretary's proposal and the argument of the Heritage author. Some may argue that the proposed transfer of functions from EPR will wrongfully continue to separate emergency management missions that many have long sought to collect in one administrative unit. They contend that the decisions to transfer preparedness functions from EPR should be reconsidered because emergency preparedness activities should be administered in proximity to the response functions to ensure that funding, technical assistance, and administrative decisions are coordinated and administered efficiently. As noted by the acting inspector general for DHS at the beginning of the 109th

Congress, he had "reservations about segregating FEMA's preparedness function from its response and recovery responsibilities. Disaster preparedness, response and recovery are integrally related, each relying on the other for success. The proposal should be studied very carefully before it is put into practice."[94] In similar fashion, a former FEMA official reportedly voiced objection to the separation of the two functions.

> Bruce Baughman, who was chosen by Vice President Dick Cheney after the 2001 attacks to head up the Office of National Preparedness, a forerunner of the Department of Homeland Security, said separating the people who plan disaster response from responders "was a big mistake. We tried that before, and it was a disaster," said Mr. Baughman, who is now the director of Alabama's Emergency Management Agency.[95]

These and others speaking against the plan disagree with the Secretary's contention that preparedness is a "distraction" from the core mission of EPR (and FEMA). In a letter to Members of Congress, the President of the National Emergency Management Association (NEMA), the professional association of state emergency management officials, argued against the proposed separation of preparedness functions from those associated with consequence management, as follows.

> These functions are closely related; response and recovery operations are based on plans created by the preparedness function. Plans are revised based on input from the response and recovery function. An effective disaster response is predicated upon planning, training and exercise. Any unnecessary separation of these functions will result in a disjointed response and adversely impact the effectiveness of departmental operations.[96]

Contrary to the Secretary's statement before Congress that "the idea here is not to decouple the skills of FEMA from preparedness," state emergency managers and others view the effect of the proposal as accomplishing just that.

Overview of Response Issues

Heroic efforts are expected after a catastrophe to meet the needs of victims and halt the effects of the disaster. Federal activities and resources are authorized to reduce the chaos, speed assistance, and coordinate the complex challenge of creating an orderly response effort. Enactment of the HSA, establishment of DHS, development of the *National Response Plan*, and other initiatives are intended to alleviate the scenario envisioned by one analyst a year before the 9/11 attacks.

> To crown this list of worries, local officials predict that long after victims of a chemical attack had been transported to hospitals, they would be bombarded with incoming federal rescue teams that would joust with each other to find something useful to do when not ordering local rescuers about in their home city. These teams, which could not arrive in time to make a lifesaving difference, would create another disaster of sorts. The list of problems deviates slightly from city to city. Although they have made headway in some areas, even cities that have benefitted from the federal unconventional terrorism preparedness programs can identify gaps in their planning and capabilities to deal with a large-scale chemical terrorist event.[97]

Despite the announced intention of Secretary Chertoff that response and recovery remain core functions of FEMA, implementation of the 2SR proposal would appear to reduce some of the agency's response authority. While details remain unknown, the proposal would have the Office of Operations Coordination (OOC) report directly to the Secretary and not to the Director of FEMA. Little information has been released on the expected authority of the OC regarding response operations. For example, the Stafford Act authorizes the President to designate a federal coordinating officer (FCO) upon the declaration of a major disaster or emergency.[98] Also, the *National Response Plan* provides that the Secretary of DHS may designate a principal federal official (PFO) to coordinate "overall federal incident management and assistance activities across the spectrum of prevention, preparedness, response, and recovery."[99] Implementation of this recommendation arguably might diminish the ability of FEMA to coordinate all facets of response operations. What relationship would the director of OOC have in relation to these coordinating officers?

The retention of response authority within FEMA could lead to a discussion of federal response capabilities and authorities that remain outside DHS. The following information is intended to help Members of Congress explore issues raised by the 2SR proposal regarding federal responsibility in three areas — emergency medical authorities, administration of the *National Response Plan*, and response authorities and resources maintained by the National Guard and Department of Defense entities. Other issues or topics will likely be identified as additional information is released by Secretary Chertoff.

Summary of Issue Discussion

Despite the positive assurances by Administration officials, implementation of the 2SR recommendation might be perceived by some to constitute a significant, and potentially negative, change in the administration of federal emergency management policies. Many, but not all, federal emergency management responsibilities have been housed in FEMA in order to build a single federal agency that would benefit from interactions and support among emergency preparedness, mitigation, response and recovery activities. In essence, FEMA is seen by some to be a synergistic environment that has enhanced federal emergency management capabilities. Some may contend that the separation of these functions from FEMA will diminish federal emergency management. In addition, the separation of preparedness assistance grant-in-aid program authority from FEMA could adversely affect efforts to build effective communications between grant administrators and staff responsible for implementing federal policy.

This concern, however, might be overstated. One might also argue that the establishment of DHS represents the most significant development in the evolution of the federal emergency management function and is consistent with past recommendations to increase the reach of federal emergency management (now homeland security) officials. Through the creation of DHS, the 107th Congress and President Bush achieved what many have sought for years — cabinet-level status based on statutory authority (not only executive directives), and formal access of emergency management officials to the President, cabinet officers, and White House staff.[100] The 2SR recommendation, some may contend, is a logical next step to be taken to ensure that federal emergency management capabilities reach the highest levels of federal governance on a continual, not intermittent, basis.[101] As summarized by one researcher in the field, "Start with the claim that disaster is normal, not special. Disasters, and even worst cases, are part of and not separate from the normal ebb and flow of social life."[102] The proposed consolidation of administrative authority in the Secretary's office may arguably be an appropriate means to manage catastrophes as a standard government function.

The decision made by Congress and President Bush to vest responsibility for comprehensive emergency management, or CEM, and the all-hazards concepts in EPR arguably provided the directorate with a strong, far reaching, congressionally mandated mission to accomplish a wide range of responsibilities. This mandate, it may be argued, is consistent with the principles first set out in 1978, an all-hazards orientation that includes all four CEM phases of preparedness, response, recovery and mitigation.

Some may contend that the 2SR proposal abrogates the CEM framework. The Secretary would appear to disagree. In testimony before the House Homeland Security Committee, the Secretary confirmed that FEMA's responsibilities are not limited by the type of event that results in a catastrophe "Clearly, FEMA has to be an all-hazards agency, and traditionally it has been the lead actor in really the core mission of making sure we are capable of responding to all hazards, including, obviously, hurricanes."[103] During the same hearing, the Secretary appeared to indicate that even if separated, the four CEM functions are bound together, as follows. "What our Preparedness Directorate will do is it will bring to the table all of these very critical functions which are part of preparedness efforts — prevention, protection and response and recovery."[104] This perspective was emphasized in a hearing held on the FY2005 budget for the directorate.[105] Support for the Secretary's position may be found in a study conducted by the Government Accountability Office (GAO). According to the GAO, the major policy initiatives undertaken by DHS and White House officials to improve the capabilities of the Nation's first responders are generally consistent with the all-hazards requirements set forth in the directives issued by President Bush.[106] The elimination of EPR and the separation of the preparedness function (and possibly mitigation) from FEMA might be considered insignificant as long as DHS ensures that policy implementation within DHS proceeds in almost seamless fashion.

Some congressional discussion ensued shortly after release of the 2SR proposal.[107] Additional debate is likely to occur. Congress might elect to debate not only the effect of the 2SR recommendation on the operations and mission of DHS, but also whether existing federal missions related to homeland security functions, including those assigned to federal entities other than DHS, might be reconsidered. The following information, and related questions, are intended to assist Members as they consider elements of that debate.

Questions and Responses

In the process of evaluating the 2SR recommendation issued by Secretary Chertoff, Congress might debate the "either-or" options — either Congress approves the Secretary's recommendation, or it does not. In the course of this debate, Members might consider the intent of Congress in creating EPR in 2002, examine whether the organization's efforts have conformed to congressional intent, and place the 2SR recommendation in the perspective of the full reach of federal homeland security related policies.

A primary question before Members of Congress is whether the 2SR proposal to eliminate EPR and modify the mission of FEMA is an appropriate response to the challenges and questions facing the nation. The history of FEMA's establishment, the struggles encountered by that new agency's administrators in the 1980s and early 1990s, and the shifts in emergency activities to and from EPR indicate that Congress and executive branch officials have long wrestled with the need to improve the administration of federal emergency authorities. Whether consolidated into FEMA in 1978, or into EPR in 2002, some authorities remained outside the jurisdiction of the unit putatively identified as responsible for "comprehensive all hazard" management. Therefore, in considering whether the 2SR proposals regarding EPR should be implemented, several broad questions and general responses might be posed in light of the "core principles" followed by DHS in preparing the 2SR initiative.

- *First principle:* "DHS must base its work on priorities driven by risk." *Question:* Does the proposal to move the remaining preparedness functions out of EPR represent a significant action that could decrease risk, or would it adversely affect the ability of DHS to perform its mission? *Response:* The proposed transfers arguably follow on similar actions that have been taken by Congress and the Administration since establishment of DHS. Critics contend that separating preparedness activities from response and recovery would inhibit feedback and interaction among emergency management units. The Secretary contends that the transfer would enable FEMA to focus on response and recovery. In light of the proposed transfer of some response functions out of FEMA, the role of other federal agencies in providing response and recovery assistance, and the need to exercise consequence management coordination at the highest levels of government, it appears that the proposed transfer would likely have some, but not a significant, impact if communication and feedback mechanisms are adopted within DHS to ensure that lessons learned from simulations and actual events are coordinated. It is not possible to assess whether this proposal would lead to a reduction or increase in risk.
- *Second principle:* "Our Department must drive improvement with a sense of urgency. Our enemy constantly changes and adapts, so we as a Department must be nimble and decisive." *Question:* Would the proposed reorganization result in a more flexible department? *Response:* The large number of DHS entities that directly report to the Secretary arguably implies a less nimble or flexible organization. On the other hand, the delegation of authority to Under Secretaries would arguably result in a more decisive, flexible, and nimble entity.
- *Third principle:* "DHS must be an effective steward of public resources." *Question:* Would the elimination of EPR and the other elements of the 2SR initiative promote priority setting, improve financial management, and facilitate the measurement of performance? *Response:* It is not possible to assess the degree to which the proposal would increase efficiency and economy. It could be argued that the centralization of preparedness authorities in one directorate might reduce duplicative assignments and redundant activities. The reorganization, however, would likely increase administrative burdens because staff charged with carrying out responsibilities in offices other than FEMA will have to increase efforts to communicate information within DHS.

- *Fourth principle:* "Effective security is built upon a network of systems that span all levels of government and the private sector." *Question:* Would implementation of the proposal facilitate interactions among governmental and non-governmental entities responsible for preparing for, easing the consequences of, responding to, and recovering from catastrophes? *Response:* The degree to which the proposed reorganization would improve the interactions of DHS with its partners outside the federal government would depend upon the ability of officials with new responsibility to build upon established relationships.[108] The identification of DHS-wide regional offices would be a factor in building these relationships.[109] Also, a large universe of federal emergency management and homeland security authorities falls outside the jurisdiction of DHS. Insufficient information exists to determine how the reorganization would affect DHS relationships with other federal agencies.

In addition to these broad questions, Members might elect to consider the following questions that more specifically address concerns associated with the proposal.

- The Homeland Security Act (HSA) directs that FEMA maintain an "all-hazards" orientation that includes terrorist attacks (except preparedness activities), natural disasters, and hazardous materials incidents. This approach to homeland security has a decades-long history involving emergency management authorities. Federal statutes other than the HSA vest certain emergency management authorities in departments or agencies other than DHS or FEMA. Should the HSA be amended to expand the responsibility of DHS for federal emergency authorities currently charged to other federal entities?
- How does the scope of all federal homeland security policies in 2005 relate to the decades-old policy of emergency management that was formulated in the 1970s? Have basic concepts, such as the division of responsibilities among federal and non-federal units of government, concepts of risk and threat, and constraints on federal authority and fiscal duty, shifted?
- Would incorporation of a definition of "homeland security" in the HSA resolve questions about the reach of the authority of DHS and its entities? Where are the boundaries among related policy areas such as "homeland defense," "law enforcement," "environmental risk and management," and "health policy?"

Preparedness Authorities Questions

Questions that might be raised by Congress specifically on the transfer of preparedness authorities to PD include the following.

- Figure 1 of this report lists five activities or responsibilities that fall within the jurisdiction of EPR's Preparedness Division. Secretary Chertoff has stated that "Other FEMA functions to be transferred include the hazardous materials training and assistance program, the chemical stockpile emergency preparedness program, the radiological emergency preparedness program and the BioShield program."[110] If the NIMS Integration Center (NIC) is transferred to the Office of Operations Coordination, will all aspects of the other four be transferred to the new Preparedness

Directorate? What functions in EPR's Preparedness Division are not proposed to be transferred to the new Preparedness Directorate?

- The Office of National Security Coordination currently reports to the Under Secretary of EPR (see Figure 1). Would this office remain within EPR or be transferred to a different DHS entity? What are the advantages and disadvantages of separating contingency planning activities from FEMA or integrating those responsibilities with other preparedness activities in the new Preparedness Directorate (PD), if established? How would the proposed realignment of preparedness functions affect the quality of federal contingency planning efforts, which have been questioned in reports and at least one congressional hearing?[111]
- Considerable attention has been given to the need to build or enhance emergency warning systems. Will Secretary Chertoff's proposal separate responsibility for terror alert and warning systems from those that focus on natural hazards? What would be the rationale for transferring authority for the Integrated Hazard Information System, or IHIS, or other warning systems out of FEMA? If responsibility for HSAS is vested in PD, how will communication between HSAS staff and the Homeland Security Operations Center (HSOC) staff (proposed to be located in the new Operations Coordination office (OC)) be facilitated?
- Hazard mitigation activities are an element of emergency preparedness in that they help communities reduce the effects of disasters before they occur. What elements of the EPR's Mitigation Division will remain with FEMA? Figure 1 identifies functions currently administered in the Mitigation Division (such as risk identification and risk assessment) that are similar to the authorities currently vested in the Under Secretary of Information Analysis and Infrastructure Protection. The duties could be undertaken in the proposed Office of Intelligence and Analysis or in the Preparedness Directorate. What criteria would be used to keep certain mitigation functions within FEMA and to transfer others?
- For years, FEMA (now EPR) has built relationships with other federal agencies, state and local governments, and private sector entities to improve federal emergency preparedness functions. For example, the emergency preparedness activities associated with events around nuclear power plants involve an array of regulations, guidelines, and public participation opportunities.[112] What specific steps will be taken by Secretary Chertoff and officials implementing the 2SR initiative to ensure that these relationships and the lessons learned from decades of involvement are not lost because of the transfer? To what extent would congressionally mandated directives related to the consolidation of DHS regional offices be affected by implementation of the 2SR recommendations?[113]
- What performance measures would be used to evaluate the effectiveness of the Preparedness Directorate's activities after the reorganization? What measures have been used to examine EPR's administration of the preparedness missions for which its staff have been responsible? To what extent did the Secretary consider such performance measures in developing the 2SR recommendation to transfer preparedness functions from EPR?

Response Authorities Questions

While FEMA operated for years under an "all-hazards" mission, the range of response authorities and their assignment to federal entities other than EPR indicate that response capabilities for some hazards rest outside the directorate. The following questions might be asked of Secretary Chertoff regarding the impact of the 2SR recommendation on federal emergency response capabilities.

- What is the relationship of NDMS to the HHS Office of Mass Casualty Planning? What would be the impact of the 2SR proposal on administration of NDMS duties and on concurrent HHS response activities and related missions of other federal entities? For example, responsibility for the transportation of NDMS resources that had previously been assigned to the Department of Defense reportedly has been transferred to DHS. Has this shift affected the delivery of NDMS services?
- Have appropriate relationships and the required mechanisms developed among the DHS, HHS, and DOD? If the 2SR recommendation is implemented and the FEMA director retains authority for response, what would be the specific duties of the Secretary of DHS and the FEMA director in coordinating the federal (and non-federal) response after the President issues a major disaster declaration under the Stafford Act?
- Should responsibility for emergency medical services be shifted to DHS from the Department of Transportation?
- Have administrative functions changed regarding DEST deployment since enactment of the HSA? To what extent does the director of EPR control the DEST unit, or does it remain a resource initiated and controlled by the FBI? What have been the positive impacts identified with the transfer of DEST to EPR? Might Congress amend the HSA provision regarding the DEST role in FEMA's response efforts?
- The bombing of the Murrah building in 1995 by domestic terrorists illustrated the dominant role of law enforcement, notably the FBI, in managing the response. Bioterrorism events will require significant involvement of HHS personnel. Hazardous material accidents will involve the Department of Transportation.[114] Should terrorist attacks from foreign aggressors lead to a catastrophe, DOD resources may likely be required due to the extent of damage and the need to respond to continuing attacks. Under these and other circumstances that will likely require considerable involvement of entities other than DHS, how will the reorganized department affect the response interactions currently guided by the *National Response Plan*?
- In light of the 2SR recommendation that a Chief Medical Officer be appointed in the new Preparedness Directorate, how will federal medical response resources be coordinated after catastrophes occur? What steps will be taken by Secretary Chertoff to ensure that DHS health response capabilities are coordinated with those of HHS?
- What steps are being taken by Secretary Chertoff to ensure that federal response capabilities outside DHS, particularly those vested in DOD, will be provided in a coordinated fashion after a catastrophe that involves significant destruction and loss of life and civil government capabilities?

- The FEMA director would report directly to the Secretary under the reorganization proposal. What responsibility would the FEMA director exercise after a catastrophic event, such as a nuclear detonation in an urban area? What would FEMA's response resources and capabilities add to the range available from other agencies? If the Office of Operations Coordination holds incident management responsibilities, what would FEMA contribute?
- Through enactment of the HSA Congress directed EPR to serve as the primary responder in "all hazards." Other federal agencies' roles provide services focused on certain types of threats, e.g., nuclear power plant incidents, hazardous material spills, or biological attacks.[115] What federal resources to be maintained by FEMA will alleviate suffering after disasters that require specialized response capabilities?
- What steps are being taken by DHS officials to ensure that plans and strategies under development within DOD are coordinated with DHS? What disagreements exist between officials of the two departments? What efforts have been taken to ensure that disagreements over roles and responsibilities are resolved?
- What performance measures will be used to evaluate the effectiveness of FEMA's response activities after the reorganization?

SUMMARY OF OPTIONS FOR CONGRESS

The 2SR process has, from the Administration's perspective, identified shortcomings that arguably require reorganizations and realignments of federal emergency preparedness and response functions. This section summarizes some of the options that might be considered by the 109th Congress as catastrophes of all kinds, and the preparation for and response to those events, become prominent aspects of federal policymaking.

Take No Legislative Action

The transfer of preparedness functions and the existence of minimally funded or supported authorities within EPR arguably serves as a justification for the recommendation of Secretary Chertoff that the directorate be reduced if not eliminated. Implementation of that recommendation, however, would require legislation since EPR was established through the HSA. By taking no legislative action, Congress would, in effect, oppose the Secretary's recommendation and maintain EPR as a directorate within DHS. Should Congress decide that the mission of EPR requires further reconsideration, resource allocation decisions made through appropriations could serve as the vehicle for change.

Place a Reconstituted FEMA in the Executive Office of the President

In the wake of the tragedy associated with the response to Hurricane Katrina and the long-term recovery issues that seem apparent, it may be argued that consequence management after certain catastrophic disasters requires action and oversight at the highest level of government. Congress could agree with the 2SR recommendation that preparedness assistance would be separated from FEMA's mission, and might consider the separation of mitigation assistance as well. The immediate and long-term coordination efforts required arguably could be vested in a FEMA director who reports directly to the President, or the Homeland Security Council. Precedent for this option may be found in President Eisenhower's reorganization plan submitted to Congress in 1958.[116]

Further Consolidate Selected Missions

The National Commission on Federal Service (referred to as the Volcker Commission) concluded that federal agencies struggle to deliver public services in the 21st century through a complex structure of agencies with similar missions.[117] Overlapping authorities, duplicative demands on resources, and unfilled gaps in communication and interagency communications all contribute to public dissatisfaction and high levels of frustration among federal workers. One solution, according to the Commission, is to reorganize federal agencies to group entities, and workers, charged with similar missions, a task arguably partly accomplished with the creation of DHS.

The 2SR recommendation may be viewed as one attempt to further consolidate those missions. If Congress is in agreement with this option, Members might elect to go beyond the reach of the 2SR recommendations by consolidating other federal emergency authorities into the jurisdiction of DHS. For example, pending legislation (H.R. 1414) would authorize the Secretary of DHS to regulate shipping of extremely hazardous material, a mission arguably suitable for the Secretary of Transportation. Other legislation (H.R. 1562) would authorize the Secretary of DHS, in consultation with the Administrator of the Environmental Protection Agency (EPA), to regulate security at chemical facilities. Another bill before Congress (S. 1256) would give the Secretary of DHS expanded regulatory authority to govern the transportation and storage of hazardous material by rail in "high threat corridors." The Department of Transportation also exercises authority over the transportation of such material.[118] Other authorities that might be transferred to DHS, or possibly EPR, authority, can be identified through an examination of Table 2 of this report.

For example, one option that Congress might consider in evaluating FEMA's emergency medical response authorities has been developed by the Homeland Security Policy Institute associated with George Washington University. Institute staff have recommended that federal responsibility for emergency medical services (EMS) should be shifted from the National Highway Traffic Safety Administration (NHTSA) in the Department of Transportation to DHS. According to the authors of the report, this recommendation should be considered along with other 2SR recommendations "to provide EMS with the leadership, resources and stature that have been absent during its recent history."[119] In response, one opponent of the proposal reportedly noted:

> The federal EMS community is not just located in DOT. There are EMS programs in the Department of Health and Human Services, DHS, and the Centers for Disease Control and Prevention. Any reorganization of federal EMS needs to look at all of these agencies in a holistic way and not simply pluck one agency from DOT and drop it down in DHS....[120]

The recommendation to transfer EMS to DHS appears to be pertinent to the Secretary's recommendation that a Chief Medical Officer be appointed within PD "to coordinate with our partners at the Department of Health and Human Services, the Department of Agriculture and state governments."[121] The Chief Medical Officer would have responsibilities beyond those normally vested in PD, arguably the coordination of federal EMS responsibilities, as follows:

> Maybe I can just touch briefly on the chief medical officer. The idea with a chief medical officer is precisely to give us somebody who owns the entirety of this system, of response with respect to health issues. That would be prevention, protection and response and recovery, because in many cases, particularly dealing with biological threats, response and recovery is a very, very important element of our defense strategy. Give that ownership to one person or one set of people and, a particular individual who I think the president has announced his intent to nominate is someone who actually has a background as an emergency room physician.[122]

While many recognize DHS accomplishments in coordinating federal emergency medical authorities, some have identified areas for improvement.[123] For example, one research center reported at the end of 2004 that responsibility for federal biodefense activity is dispersed among "more than a dozen government agencies." The report concluded that the existing array of authorities "presents particular challenges to efforts to design, implement, and oversee a coherent, coordinated, and efficient biodefense strategy."[124] Also, a witness at a Senate hearing at the beginning of the 109th Congress noted that federal bioterrorism response strategy needs "fine tuning" and added, "I'm not really sure DHS needs a role in BioShield at all."[125] The dominant role of HHS through the Centers for Disease Control and Prevention (CDC) was summarized by the CDC director as follows.

> HHS is responsible for leading federal public health efforts to ensure an integrated and focused national effort to anticipate and respond to emerging biological and other weapons threats. HHS is also the principal federal agency responsible for coordinating all federal-level assets activated to support and augment the state and local medical and public health response to mass casualty events.[126]

These statements raise questions about which department, DHS or HHS, has primary coordination authority. The issue of coordinating federal activities in biomedical defense has been explored by a House subcommittee as well.[127] Further information is required to determine whether the 2SR recommendation addresses this issue.

As part of the process of evaluating whether certain emergency authorities should be transferred to DHS, Members might consider the following questions in developing criteria for legislation that would further consolidate authorities in DHS.

- Does the function require considerable or ongoing coordination between DHS and other federal entities, not just at the time a catastrophe strikes and FEMA's response and recovery mission is needed?
- Does the function require technical knowledge, skills, or resources that are an inherent part of another agency's or department's mission? Would the "generalist" emergency management skills of FEMA or other DHS officials have to be

supplemented with the technical skills that are part of the other department or agency base mission?

- Would the incorporation of the function into the mission of DHS reduce the number, and therefore the complexity, of federal agencies with responsibility for similar activities?
- Do the non-federal entities involved in providing services associated with the function have long-standing connections with federal entities other than DHS? How would the transfer to DHS affect the ongoing review of regional offices?

Another approach to consolidating these authorities would be to expand legislation before the 109th Congress (H.R. 1817) already approved by the House. This bill, which would authorize appropriations for DHS, would require the preparation of a "Terrorism Prevention Plan" and amend the Homeland Security Act (HSA) to establish a "National Terrorism Exercise Program." Title III of the bill would require that such a program be established in ODP in order to test and evaluate domestic capabilities to "prevent, prepare for, respond to, and recover from threatened or actual acts of terrorism...." Congress might elect to debate whether this proposal might be expanded. It might be argued that the National Terrorism Exercise Program should include all hazards, not just terrorism. This would be consistent with the overall mission given to DHS in the HSA[128] and with the all-hazards preparedness guidance and national preparedness system released by DHS in 2004 and 2005.[129] If Members contemplate expanding the program to an all-hazards effort, it might be argued that EPR, rather than ODP, would be the appropriate administrative home.

Continuity of Operations

It might be argued that federal continuity of operations (COOP) activities might be improved if EPR's Office of National Security Coordination were given greater authority to guide agencies that must improve their own internal planning efforts. Congress might consider the need for statutory authority that would specify the relationship of DHS efforts to those of other agencies. For example, while DHS (specifically FEMA) has been designated the lead agency to establish guidelines and standards for federal agency COOP efforts, it may not be authorized "to compel action by other executive branch agencies."[130] Legislation might be considered to authorize EPR to exercise strong coordinative functions that would improve COOP efforts. On the other hand, opponents of such strong central authority within DHS might contend that guidance and technical assistance, not mandatory enforcement, are the appropriate roles for EPR. Others might argue that the essential function of ensuring the continuity of operations should be vested in the Secretary's office or the White House.

Hazard Warning Systems

The tsunami that struck Southeast Asia on December 26, 2004, has raised awareness throughout the world of the need for monitoring systems. During the 108th Congress, legislation was introduced (H.R. 3644) to establish a technology transfer program within DHS to improve the capabilities of emergency responders to counter terrorist and non-terrorist threats. One news report indicates that the issue may be raised before the 109th Congress.[131] Debate over the future of IHIS could involve the broader discussion of the need for hazard warning systems. In addition, Congress might consider whether the capabilities of another federal agency should be augmented. For example, the Domestic Warning Center

(DWC) is used by the Department of Defense (NORTHCOM) to track disasters.[132] Legislation might be considered to incorporate these resources into DHS.

Separate Natural Disaster and Terrorism Missions

One study on initiatives that, according to the authors, transformed the federal government identified two lessons learned from the revitalization of FEMA during the 1990s — clarify the mission of the agency, and structure the agency to reflect the mission.[133] The authors noted that the decision to transfer the focus of FEMA from national preparedness to emergency management "redefined the agency's primary target population as disaster victims, rather than executive branch officials central to the survivability of national decision-making capacity following a nuclear war."[134]

Members of Congress might elect to consider whether the current emphasis on terrorist threats dilutes the FEMA mission or broadens it without harming its "focus," whether viewed as "emergency management," "natural disaster assistance," or "all-hazards." If, as the authors of the study noted, "national preparedness functions were not abandoned, but were integrated with the more basic emergency management functions," one might argue that returning to a more focused mission (i.e., natural disaster preparedness and response) rather than "all-hazards" approach, would be more appropriate. This might be accomplished by eliminating EPR as a separate DHS entity, restoring FEMA to its independent agency status, or distributing "natural disaster" emergency management functions among other federal agencies.

Strengthen and Monitor Interagency Coordination Requirements

Congress might encourage or require changes in interagency coordination mechanisms. In the immediate aftermath of the terrorist attacks of 2001, President Bush established the Homeland Security Council (HSC) to "ensure coordination of all homeland security-related activities among executive departments and agencies and promote the effective development and implementations of all homeland security policies."[135] Homeland security policy issues are evaluated by the Homeland Security Council Deputies Committee, which includes the director of FEMA. Eleven policy coordination committees (PCCs) "coordinate the development and implementation of homeland security policies by multiple departments and agencies throughout the federal government...." Two PCCs particularly appropriate to the mission of EPR include one on "Plans, Training, Exercises, and Evaluation" and another on "Domestic Threat Response and Incident Management." Members of Congress might conduct hearings on the role of these PCCs to evaluate whether the task assigned in the presidential directive is being carried out, and whether other means of coordination would be more appropriate.[136]

One possible model for this approach is the establishment of joint command responsibilities in DOD through enactment of the Goldwater-Nichols Department of Defense Reorganization Act of 1986.[137] This statute, according to one analyst, addressed problems associated with the imbalance among three "basic military organizational tensions:

centralization versus decentralization, functional versus geographic responsibility, and specialization versus the generalist perspective."[138]

Two of these three (centralization/decentralization and specialization/generalist) are applicable to the complex array of federal authorities administered by DHS and other departments and agencies. Congress might consider two options in adapting the Goldwater-Nichols model to federal homeland security policy. First, the statutory authority of the Secretary of DHS could be expanded (Section 102 of P.L. 107-296) to mandate negotiations between the Secretary and other executive agency heads over the use of federal resources under specified emergency conditions or threats. Such authority could enable the Secretary to supplement the generalist skills within DHS with the technical skills and resources of other agencies.

Second, legislation might be considered to authorize specified officials under certain circumstances to call on the resources of other federal agencies in coordination with the Secretary of DHS or the Homeland Security Advisory Council (HSAC). A "Federal Emergency Response Coordinator" from DHS, HHS, or another federal office might be tasked with ensuring, on a continuing basis, that federal emergency preparedness and response actions are coordinated and complementary, not duplicative.[139] Under Homeland Security Presidential Directive-5, the DHS Secretary "is responsible for coordinating federal operations within the United States to prepare for, respond to, and recover from terrorist attacks, major disasters, and other emergencies." Congress might elect to build upon this administrative authority of the Secretary by amending the Homeland Security Act to require that certain actions be taken, such as the convening of department heads on a regular basis or the preparation of an emergency response capabilities report for congressional evaluation.[140] Members might also consider attaching coordination requirements or additional resources to departments' annual appropriations legislation. Another variant on this options would be to amend Title IX of the HSA to authorize the HSAC to require federal agency coordination and intervene to resolve disputes among department heads, including Secretary of DHS.

Other coordination options might be considered by Congress. A "virtual" reorganization through coordinating councils has been suggested by one authority, who concluded in a study of federal reorganizations that an emphasis on the process used to coordinate the actions of federal agencies can result in improvements previously associated only with reorganizations.[141] Enactment of statutory provisions requiring such coordination could specify reporting requirements to congressional committees with jurisdiction, the integration of agency strategic plans required under the Government Performance and Results Act, and the identification of an agency head as the chair of the interagency effort.

Finally, the Policy Directorate that Secretary Chertoff has proposed in the 2SR recommendation could serve an important role in ensuring that DHS-wide activities are coordinated, and that the administrative components of the department, including FEMA, undertake specified tasks to ensure that stovepipes within DHS are either circumvented or porous to facilitate communications and information sharing within the department.

CONCLUSION

The Homeland Security Act of 2002 established the statutory framework for DHS and its components, including EPR. From enactment of the HSA in 2002 to the present, changes have taken place with regard to the mission of EPR. Elements identified in the HSA as components of EPR's comprehensive emergency management responsibility have

- been excluded from the mission of EPR,
- received little or no support when transferred,
- been subsequently transferred from EPR, or
- been assigned to other entities.

The Administration's 2SR recommendation to eliminate EPR and reduce FEMA's responsibilities is consistent with past trends. Because FEMA remains in DHS, however, it may be argued that the changes are not significant because DHS as a whole exercises broad emergency authorities.

Members of the 109th Congress might elect to approve the Administration's 2SR proposal to eliminate EPR. Congress might also consider consolidating emergency authorities vested in other federal departments and agencies into DHS or reduce the need to reorganize DHS by enhancing interagency coordination requirements. Congress might elect to examine the criteria used to justify the transfer of emergency management functions from EPR to other federal entities.

APPENDIX A: ACRONYM GLOSSARY

Acronym	Meaning
2SR	Second stage review
CDC	Centers for Disease Control and Prevention
CEM	Comprehensive emergency management
COG	Continuity of government
CONPLAN	Concept of Operations Plan
COOP	Continuity of operations
DESTs	Domestic Emergency Support Teams
DHS	Department of Homeland Security
DOD	Department of Defense
DOT	Department of Transportation
DWC	Domestic Warning Center
EMS	Emergency medical services
EPR	Emergency Preparedness and Response Directorate
FBI	Federal Bureau of Investigation
FEMA	Federal Emergency Management Agency
FICEMS	Federal Interagency Committee on Emergency Medical Services
FIRSTs	Federal Initial Response Support Teams
FPC	Federal Preparedness Circular
HAZMAT	Hazardous material
HHS	Department of Health and Human Services
HSA	Homeland Security Act
HSAS	Homeland Security Advisory System

HSC	Homeland Security Council
HSPD	Homeland Security Presidential Directive
IAIP	Information Analysis and Infrastructure Protection
IEMS	Integrated emergency management system
IHIS	Integrated Hazard Information System (also referred to as FIRESAT)
IMAAC	Interagency Modeling and Atmospheric Assessment Center
MMRS	Metropolitan Medical Response System
NDMS	National Disaster Medical System
NDPO	National Domestic Preparedness Office
NHSA	National Homeland Security Agency
NHTSA	National Highway Traffic Safety Administration
NIC	NIMS Incident Center
NIMS	National Incident Management System
NIRT	Nuclear Incident Response Team
NOAA	National Oceanic and Atmospheric Administration
NORTHCOM	Northern Command
NRP	National Response Plan
ODP	Office for Domestic Preparedness
OEP	Office of Emergency Preparedness
ONSC	Office of National Security Coordination
PCCs	Policy coordination committees
PHSA	Public Health Service Act
SAC	Special Agent-in-Charge
SLGCP	Office for State and Local Government Coordination and Preparedness
SNS	Strategic National Stockpile
WMD	Weapons of mass destruction

Appendix B: Evolution of Federal Emergency Authorities

From the early years of the republic to 1950, Congress enacted legislation that directed federal disaster relief. Laws that were unique to each disaster authorized the amount of funds to be distributed, the type of federal equipment to be sent, or the personnel to be allocated to stricken areas.[142] For the most part, federal emergency assistance consisted of disaster relief authorized to provide specific relief to disaster victims.[143]

In 1950, Congress and President Truman approved legislation that authorized the President to determine when federal assistance would be provided, subject to stated eligibility constraints.[144] Following enactment of the 1950 statute, administrative responsibility for the provision of federal emergency assistance (as well as civil defense, also authorized in 1950) shifted among federal departments, agencies, and the White House until 1978[145] During that time, the provisions of the 1950 statute were expanded upon and superseded through significant legislation enacted from 1966 to 1974 that increased the categories of assistance to be provided and the types of organizations eligible for aid.[146] In addition, other legislation was enacted to improve civil defense efforts; reduce future disaster losses (hazard mitigation); improve safety from nuclear and industrial catastrophes; and provide relief for small businesses, agricultural producers, and homeowners. All of these authorities developed into a complex mix of federal emergency management missions that, by the late 1970s, involved many federal agencies.

State, local, and federal officials increasingly voiced dissatisfaction over the complex and inefficient maze of federal policies and the responsible administrative entities. The reorganization project headed by OMB during the Carter Administration associated this conundrum with the decision made by President Nixon in 1973 to disperse authorities out of the Executive Office of the President, summarized as follows.

> This reorganization also meant that all three of the major agencies concerned with civil emergency preparedness after June 30, 1973, maintained their own separate regional offices. Consequently, state officials were required to deal with at least three sets of federal regional officials on often closely related substantive program issues. It is probably fair to say that state and local dissatisfaction with the fragmentation of federal emergency preparedness organizational arrangements grew substantially after the 1973 reorganization.[147]

Through a series of discussions and studies OMB officials developed a new policy framework that consolidated emergency management functions into a four-part policy framework comprising preparedness, response, recovery, and mitigation missions. This amalgam of policies was referred to as comprehensive emergency management (CEM).[148]

Building upon the work of an interagency reorganization team and using the CEM framework, President Carter proposed that certain emergency operational and coordination functions previously dispersed throughout the federal government be brought under the jurisdiction of one independent agency. Administration officials recommended that a new federal entity be established to administer many of the federal policies related to the management of emergencies. Through a reorganization plan submitted to Congress in 1978, President Carter advocated the establishment of FEMA based on the following four fundamental principles.

- First, federal authorities to anticipate, prepare for, and respond to major civil emergencies should be supervised by one official responsible to the President and given attention by other officials at the highest levels.
- Second, an effective civil defense system requires the most efficient use of all available emergency resources.
- Third, whenever possible, emergency responsibilities should be extensions of the regular missions of federal agencies.
- Fourth, federal hazard mitigation activities should be closely linked with emergency preparedness and response functions.[149]

Under the second principle, an all hazards approach was perceived to be needed to ensure that civil defense resources and systems for warning, evacuation, and preparedness would be available for any disaster, regardless of cause, in order to achieve the "most efficient use of all available emergency resources." The all-hazards principle advocated by the Carter Administration paralleled legislation previously approved by Congress[150] Equipment, plans, procedures, and policies needed to prepare for and respond to one type of catastrophe (e.g. a natural disaster) can, under the all-hazards concept, be applied to emergency management tasks associated with other catastrophes (such as terrorist attacks).[151]

The third reorganization principle enunciated in the reorganization plan advocated the retention of emergency responsibilities in agencies as "extensions" of their regular missions. As detailed in the plan, this would be achieved by authorizing FEMA "to *coordinate and plan* [emphasis added] for the emergency deployment of resources that have other routine uses. There is no need to develop a separate set of federal skills and capabilities for those rare occasions when catastrophe occurs."[152] In summary, the Office of Management and Budget study that led to the creation of FEMA advocated the centralization of responsibilities within the new agency, but perceived the need to spread operational responsibilities across the government to regulate and manage specific types of hazards.[153]

Commensurate with the third principle (and with the absence of legislative action), some emergency management functions were not transferred to FEMA in 1978. For example, disaster loans for small businesses and agricultural producers continued to be administered by the Small Business Administration and the Department of Agriculture. Hazardous materials and oil spills remained part of the preparedness and response missions of the Environmental Protection Agency and the U.S. Coast Guard. Rather than build redundant programs and activities in the new agency, the Carter Administration and Congress left unchanged the missions of many agencies.

The task of centralizing some emergency authority in FEMA resolved some problems; however, the retention of parallel authorities elsewhere complicated efforts by the new leadership. For more than a decade after the agency's establishment, FEMA officials struggled to develop the agency and bring coherence to the range of federal authorities and missions viewed by many to be muddled and inefficient. The task of coordinating these activities was formidable, due in part to the essential role exercised by state and local governments as well as non-governmental organizations.[154] Some of FEMA's federal "partners" did not readily cede authority, and even units within the agency reportedly showed signs of poor cooperation. The low point may have been reached in the late 1980s and in 1992 after critics claimed that the responses to Hurricanes Hugo (1988) and Andrew (1992) were faulty. Some called for FEMA's abolition. Other legislation would have significantly reformed the agency.[155] In the end, significant administrative changes, not legislative, were made in the 1990s and guided FEMA to a reputation amongst some as a model agency.[156] Issues that challenged FEMA's leadership over the first decade of the agency's existence included the following:

- inconsistent access to the President and lack of administrative strength to force other federal administrators to share authority, as the FEMA director was not a member of the Cabinet and headed a relatively small independent agency with a limited budget;
- he integration of different cultures and missions (fire safety, civil defense, insurance, environmental management) into one agency with a broad, new, and undefined policy agenda;
- the seemingly far-reaching authority in policy areas (comprehensive emergency management for all hazards) that touched upon the missions of other federal agencies and departments.

One explanation, among others, for the struggles FEMA administrators encountered to accomplish its missions rests with the discrepant second and third principles set out in President Carter's 1978 reorganization plan. As originally envisioned, FEMA officials were expected to meet the full range of emergency management needs associated with the CEM framework, for all hazards. This is arguably the broadest mandate possible, which might cover events before and after incidents from complex national security threats, industrial accidents that could threaten hundreds or thousands of lives, nuclear power plant malfunctions, and public health threats and emergencies, to relatively minor incidents such as storms that resulted in some damages and losses. The mission of FEMA was potentially as broad as any federal policy area. FEMA progressed in wobbly steps, due to the difficulties of balancing the two potentially conflicting principles of administering its comprehensive all-hazards mission and meeting the visible and public challenges of helping communities stricken by catastrophes, while being constrained by parallel or convergent authorities of other federal agencies. In short, the theoretically wide-ranging emergency management policy mission for FEMA had to be squeezed between the "regular missions" of other federal agencies.

One might contend that the success of the agency in the 1990s was due, in large part, to the focus on natural disasters and the limited need to be concerned with other threats, such as those stemming from terrorist attacks or industrial accidents. The complexity of the latter types of incidents, arguably might have challenged the abilities of the small independent agency.[157]

Some recognized that FEMA's focus on natural disasters, and the dispersed authority for aspects of emergency management among federal agencies, created gaps in emergency management capabilities. For example, in the winter of 2001 the U.S. Commission on National Security/21st Century, also known as the Hart/Rudman Commission, reported on the lack of preparedness for catastrophic terrorist attacks involving weapons of mass destruction. The Commission recommended establishment of a National Homeland Security Agency (NHSA) based largely on FEMA's all-hazards mission in order to better coordinate federal policy.[158] Also, the perspective of state and local officials who sought preparedness assistance from federal agencies was summarized by one analyst roughly a year before the 9/11 attacks as a process of inter-agency jockeying. This complaint appears reminiscent of complaints raised in the 1970s that led to the establishment of FEMA.

> For their part, local officials soon deduced that the federal "partners" were busier competing with each other for missions and resources than they were coordinating their efforts. The equipment grant programs of the Defense, Health and Human Services, and Justice Departments all had varying timelines and requirements, slightly different goals, and conflicting views on priorities regarding how to accomplish cerain response tasks. Another byproduct of the lack of federal coordination was the creation of roughly ninety terrorism preparedness courses. Firefighters alone could get training from three federal agencies, headlined by the Army's Domestic Preparedness Program.[159]

The terrorist attacks of September 11, 2001, and perhaps more even more significantly, the tragedy associated with Hurricane Katrina have refocused attention on the deficiencies in federal emergency management policies and administrative functions. While not agreeing with all of the Commission's recommendations, Congress acted upon some of it's findings following the 9/11 attacks. Enactment of the Homeland Security Act of 2002 (HSA) established the Department of Homeland Security (DHS). Congress vested the new department with greater authority and standing than envisioned by the Hart/Rudman Commission for NHSA. The statute consolidated authorities beyond those given to FEMA in 1978 and created a place for FEMA inside the new department. Title V of the HSA established the Emergency Preparedness and Response (EPR) Directorate and retained the FEMA mission as one of all-hazards comprehensive emergency management, except for terrorism preparedness, responsibility for which was given to the Border and Transportation Security Directorate (Title IV of the HSA).

In summary, the evolution of the emergency management policy area into the new, and undefined, homeland security policy field, and the transition of the organizations from FEMA to DHS mirror past actions and issues of debate. Secretary Chertoff's 2SR proposal is the next step in the evolution of the policy area and the assignment of duties to the appropriate administrative entity. Just as President Carter, with the consent of the 95th Congress, consolidated missions into FEMA in 1979, President Bush and the 107th Congress passed the HSA in 2002 to more fully integrate and coordinate federal emergency preparedness and response missions.[160] However, the authorities that brought FEMA into existence in 1979 did not consolidate all federal emergency management authorities, and, as a result, the reorganization resulted in an imbalance between the perception that FEMA could exercise broad authority and the reality of constraints faced by agency officials. Similarly, the HSA did not centralize all federal emergency functions within FEMA (or the EPR directorate), and also did not consolidate all pertinent authorities into DHS. The imbalance remains as FEMA and other DHS officials continue to struggle to establish cooperative mechanisms that bridge jurisdictional boundaries. The 2SR proposal is an attempt to create greater balance. Members of the 109th Congress and Administration officials continue to wrestle with the dilemma of matching emergency (homeland security) policy and missions to organizations.

ENDNOTES

1 "Statement by Secretary of Homeland Security Michael Chertoff before the House Appropriations Homeland Security Subcommittee," Mar. 2, 2005, available at [http://www.dhs.gov/dhspublic/display?theme=45&content=4381], visited June 14, 2005. The findings are summarized on the DHS website, "Secretary Michael Chertoff U.S. Department of Homeland Security Second Stage Review Remarks," available at [http://www.dhs.gov/dhspublic/display?content=4597], visited July 13, 2005. Secretary Chertoff has explained that the establishment of DHS as a cabinet-level department under then Secretary Ridge constituted the "first stage"; Secretary Chertoff's proposal is presented as the "second stage."

2 The Administration submitted an FY2006 budget amendment to Congress in order to implement the changes at the beginning of the new fiscal year. See the July 22, 2005, letter from the director of the Office of Management and Budget at [http://www.whitehouse.gov/omb/budget/amendments/amendment_7_22_05.pdf], visited Aug. 1, 2005. For an overview, see CRS Report RL33042, *Department of Homeland Security Reorganization: The 2SR Initiative*, by Harold C. Relyea and Henry B. Hogue.

3 Jim Morris, "Chertoff Says First Phase of Sweeping DHS Review Finished Ahead of Schedule," *CQ Homeland Security*, June 9, 2005.

[4] James Jay Carafano and David Heyman, *DHS2.0: Rethinking the Department of Homeland Security* (Washington: The Heritage Foundation, 2004).

[5] Ibid, p. 14.

[6] Response of Secretary Chertoff before House Committee on Homeland Security, *Review of Department of Homeland Security Organization* (Washington: 2005), July 25, 2005, transcript available by subscription through CQ Homeland Security.

[7] U.S. Department of Homeland Security, Secretary Michael Chertoff, letter to Congress of July 13, 2005.

[8] Sec. 501, P.L. 107-296, 6 U.S.C. 311.

[9] 6 U.S.C. 313(1)-(6).

[10] Appendix B of this report provides background information. Many publications cover the establishment of FEMA in 1978 and the evolution of its mission over the years. See, for example, Richard Sylves and William R. Cumming, "FEMA's Path to Homeland Security: 199-2003," *Journal of Homeland Security and Emergency Management*, vol. 1, issue 2, 2004, article 11. National Academy of Public Administration, *Coping with Catastrophe: Building an Emergency Management System to Meet People's Needs in Natural and Manmade Disasters* (Washington: 1993).

[11] The act renames the IHIS system "FIRESAT." Funding for this program has not been authorized since FY2000. The House report that accompanied the HSA legislation (H.R. 5005) noted that IHIS would give DHS "a real near-time capability to detect wild fires in North America."

[12] According to the FY2003 budget request submitted by the Federal Bureau of Investigation (FBI), no funds were requested as the "NDPO consequence management activities will be transferred to the Federal Emergency Management Administration's [Agency] new terrorism office. However, the FBI will continue to retain responsibility for crisis management." See "Department of Justice FY2003 Budget Summary, Federal Bureau of Investigation Salaries and Expenses," at [http://www.usdoj.gov/jmd/2003summary/html/fbi-se-bs2.htm], visited Feb. 8, 2005. Scant information exists on the extent to which the NDPO is used, funded, or considered a resource.

[13] DEST is a stand-by interagency team of experts that provides an on-scene commander (Special Agent in Charge) with advice and guidance in situations involving a weapon of mass destruction (WMD) or other significant domestic threat. More information on the role of DEST in the response activities of EPR appears later in this report.

[14] The HHS components transferred to EPR in March 2003 included the National Disaster Medical System (NDMS). For summary information, see U.S. Department of Homeland Security, "National Disaster Medical System," at [http://www.fema.gov/news/newsrelease_print.fema?id=11927], visited Apr. 22, 2005. For related information on federal health authorities, see CRS Report RL31719, *An Overview of the U.S. Public Health System in the Context of Emergency Preparedness*, by Sarah A. Lister.

[15] The SNS is no longer part of the EPR mission. The Project Bioshield Act of 2004 (Sec. 3, P.L. 108-276) authorizes the Secretary of HHS, "in coordination with the Secretary of Homeland Security," to administer the SNS.

[16] Sec. 504 and Sec. 506, P.L. 107-296, 6 U.S.C. 314, 316.

[17] This provision was amended in the Intelligence Reform and Terrorism Prevention Act of 2004, P.L. 108-458.

[18] Sec. 502, 505-509, P.L. 107-296, 6 U.S.C. 312, 315-319.

[19] 6 U.S.C. 317.

[20] "Department of Homeland Security Organization Chart," available at [http://www.dhs.gov /interweb/assetlibrary/DHSOrgCharts0705.pdf], visited July 13, 2005.

[21] Sec. 430 of P.L. 107-296, 6 U.S.C. 238.

[22] "The Secretary may allocate or reallocate functions among the officers of the Department, and may establish, consolidate, alter, or discontinue organizational units within the Department, but only (1) pursuant to section 1502(b); or (2) after the expiration of 60 days after providing notice of such action to the appropriate congressional committees, which shall include an explanation of the rationale for the action." Sec. 872 of P.L. 107-296, 6 U.S.C. 452. The reorganization was proposed in Secretary Ridge, letter to Senator Susan Collins, Chair, Senate Governmental Affairs Committee, Jan. 26, 2004. For background on the reorganization authority, see CRS Report RS21450, *Homeland Security: Scope of the Secretary's Reorganization Authority*, by Stephen R. Viña.

[23] U.S. Congress, Conference Committees, *Making Appropriations for the Department of Homeland Security for the Fiscal Year Ending September 30, 2005, and for Other Purposes*, conference report to accompany H.R. 4567, H.Rept. 108-774, 108th Cong., 2nd sess. (Washington: GPO, 2004), p. 62.

[24] P.L. 108-360, 118 Stat. 1671.

[25] Sec. 7303(h) of P.L. 108-458, 118 Stat. 3846-3847.

[26] R. Eric Petersen of the Government and Finance Division, CRS, contributed to this section.

[27] For information on COOP activities, see CRS Report RL32752, *Continuity of Operations (COOP) in the Executive Branch: Issues in the 109th Congress*, and CRS Report RL31857, *Executive Branch Continuity of Operations (COOP): An Overview*, by R. Eric Petersen.

[28] "Each department and agency shall support interagency coordination to improve preparedness and response to a national security emergency and shall develop and maintain decentralized capabilities wherever feasible and appropriate.... Emergency plans and programs, and an appropriate state of readiness, including organizational infrastructure, shall be developed as an integral part of the continuing activities of each federal department and agency." U.S. President (Reagan), "Assignment of Emergency Preparedness Responsibilities," Executive Order 12656, *Federal Register*, vol 53, Nov. 18, 1988, p. 47493.

[29] Ibid., Part 17.

[30] Statement of Reynold N. Hoover, Director, Office of National Security Coordination, Department of Homeland Security, before the Senate Committee on Commerce, Science and Transportation, Subcommittee on Disaster Prevention and Prediction, hearing on "All Hazards Alert Systems," July 27, 2 005, available at [http://commerce.senate.gov/ hearings/witnesslist.cfm?id=1591], visited Aug. 1, 2005.

[31] U.S. Department of Homeland Security, Federal Emergency Management Agency, *Federal Preparedness Circular (FPC 65)* (Washington: 2004), available at [http://www.fema.gov/onsc/docs/fpc_65.pdf], visited May 20, 2005.

[32] Ibid., p. 3.

[33] For a discussion of agency plans and authorities, see CRS Report RL31857, *Executive Branch Continuity of Operations (COOP): An Overview*, by R. Eric Petersen.

[34] Sec. 201(d)(7) of P.L. 107-296, 6 U.S.C. 121(d)(7). The HSAS was first established in U.S. President (Bush), "Homeland Security Presidential Directive-3," Mar. 11, 2002.

[35] A comment made by Senator Lieberman during a recent hearing is notable in this regard: "Nor can we tolerate a department where the officials responsible for overseeing and managing don't have adequate resources at their disposal to get the job done, because if we give them authority but not resources to get the job done, we're still setting them up for failure. And their failure, of course, is at our peril." *Outlook for the Department of Homeland Security*, Jan. 26, 2005. In evaluating the need to support FIRESAT, Congress might elect to consider the hazards threat information available through the Interagency Modeling and Atmospheric Assessment Center (IMAAC). As specified in the *National Response Plan*, the Center serves as "the single source" on the dispersion of hazardous releases in the atmosphere.

[36] Background on the HSAS and issues associated with the system are discussed in CRS Report RL32023, *Homeland Security Advisory System: Possible Issues for Congressional Oversight*, by Shawn Reese. See also U.S. Congress, House Committee on Government Reform, Subcommittee on National Security, Emerging Threats, and International Relations, *Homeland Security Advisory System: Threat Codes and Public Responses*, hearing, 108th Cong., 2nd sess., Mar. 16, 2004 (Washington: 2004).

[37] Two bills have been approved by the Senate. S. 50 would enhance the tsunami warning system administered by the National Oceanic and Atmospheric Administration (NOAA) in the Department of Commerce. S. 361 would authorize NOAA to serve as the lead federal agency for an ocean and coastal observation system.

[38] Certain elements of the response to the four Florida hurricanes in the fall of 2004 raised concerns about the process used by FEMA to contract out damage assessment inspectors and allocate funding. In addition, some have questioned decisions made in the distribution of assistance in Florida after the four hurricanes of 2004. See U.S. Senate, Homeland Security and Governmental Affairs Committee, *FEMA's Response to the 2004 Hurricanes in Florida: Were There Adequate Safeguards Against Waste, Fraud, and Abuse?*, hearing May 18, 2005, 109th Cong., 1st sess. Questions have also be raised with regard to the assistance provided to victims of Hurricane Isabel. See U.S. Congress, House Committee on Transportation and Infrastructure, Subcommittee on Economic Development, Public Buildings and Emergency Management, *The Federal Response to Hurricane Isabel*, hearing, 108th Cong., 1st sess., Oct. 7, 2003 (Washington: GPO, 2004).

[39] U.S. Department of Homeland Security, *The National Response Plan*, available at [http://www.dhs.gov/dhspublic/interapp/editorial/editorial_0566.xml], visited Feb. 1, 2005. For background on the NRP see CRS Report RL32803, *The National Preparedness System: Issues in the 109th Congress*, by Keith Bea. The term "Incident of National Significance" is defined in the NRP (see p. 67).

[40] 42 U.S.C. 5121 et seq.

[41] The NRP defines an "Incident of National Significance" as an "actual or potential high-impact event that requires a coordinated and effective response by an appropriate combination of federal, state, local, tribal, nongovernmental, and/or private-sector entities in order to save lives and minimize damage, and provide the basis for long-term community recovery and mitigation activities. See U.S. Department of Homeland Security, *National Response Plan*, p. 67.

[42] Sec. 502, P.L. 107-296, 6 U.S.C. 312.

[43] For example, see Eileen Sullivan, "When Building Anti-Terror Plans, Success Is Only Option," *Federal Times*, May 17, 2004, p. 22. Martin Edwin Andersen, "Local Officials Howl at DHS Emergency Management Plan," *CQ Homeland Security*, Aug. 8, 2003.

[44] As noted on page 8 of this report few resources appear to have been allocated to, and previously little authority vested in, DEST or NDPO activities.

[45] U.S. Department of Homeland Security, *National Response Plan*, p. 34.

[46] *CONPLAN United States Government Interagency Domestic Terrorism Concept of Operations Plan* (Washington: 2001), p. 3.

[47] Ibid., p. 28.

[48] 6 U.S.C. 312(3). As noted elsewhere in this report, authority over the stockpile has shifted back to HHS.

[49] U.S. President (Bush), "Management of Domestic Incidents," *Homeland Security Presidential Directive-5*, Feb. 28, 2003.

[50] U.S. Department of Homeland Security, *Secretary Michael Chertoff U.S. Department of Homeland Security Second Stage Review Remarks* (Washington: 2005).

[51] One writer has opined that "a sea of blue tarps on houses with missing roofs will remain as FEMA's response signature." See SEMP Biot #241, *The Incredible Shrinking FEMA*, available at [http://www.semp.us/biots/biot_241.html], visited Aug. 5, 2005. For a discussion of FEMA's efforts at long-term recovery in certain communities, see Denise Kersten, "Out of the Ruins," *Government Executive*, available at [http://www.gov exec.com/features/0705-15/0705-15s3.htm], visited Aug. 22, 2005.

[52] Title V of P.L. 93-288, the Disaster Relief Act of 1974 (88 Stat. 160-163) authorized the President to provide economic recovery assistance "after the period of emergency aid and replacement of essential facilities and services." Congress never funded this authority and it was repealed in the 1998 amendments (see Sec. 102(c) of P.L. 105-393, 112 Stat. 3617). The Secretary of Commerce, however, is authorized to undertake disaster economic recovery activities. See 42 U.S.C. 3149(c)(2).

[53] Sec. 404 of the Stafford Act, 42 U.S.C. 5170c.

[54] 42 U.S.C. 5133.

[55] 42 U.S.C. 4104c.

[56] U.S. President (Bush), "Critical Infrastructure Identification, Prioritization, and Protection," *Homeland Security Presidential Directive/HSPD-7* (Washington: 2003), Sec. 18.

[57] Database search assistance provided by Thomas Carr of the Government and Finance Division, CRS. In addition to the authorities listed in Table 1, other statutory provisions have a bearing on an examination of federal emergency authorities. For example, liability provisions for oil spill disasters associated with "Acts of God" are set out in 33 U.S.C. 2701 et seq. Rulemaking provisions of the Administrative Procedures Act, 5 U.S.C. 533, may be found inapplicable under emergency situations, pursuant to court rulings reported in the Notes section. The information in Table 1 excludes emergencies solely pertinent to war-related conditions.

[58] See CRS Report RL32920, *Federal Counter-Terrorism Training: Issues for Congressional Oversight*, coordinated by Shawn Reese.

[59] U.S. Department of Transportation, "Organization and Delegation of Powers and Duties; Office of Intelligence, Security, and Emergency Response," *Federal Register*, vol. 70, Feb. 15, 2005, pp. 7669-7670.

[60] U.S. Department of Health and Human Services, "Office of Public Health Emergency Preparedness Statement of Organization, Functions, and Delegations of Authority," *Federal Register*, vol. 70, Feb. 1, 2005, pp. 5183-5184.

[61] Ibid, p. 5183.

[62] 49 U.S.C. 5103.

[63] An advisory bulletin issued in the aftermath of fires and explosions that occurred in 1998 and 2003 addressed the "need for operators to plan with utilities on how to coordinate actions needed in responding to a pipeline emergency." See U.S. Department of Transportation, Pipeline and Hazardous Materials Safety Administration, "Pipeline Safety: Planning for Coordination of Emergency Response to Pipeline Emergencies," *Federal Register*, vol. 70, no. 98, May 23, 2005, p. 29557. The pertinent regulation is found at 49 CFR 195.402(e), 403. For information on the national preparedness system, see CRS Report RL32803, *The National Preparedness System: Issues in the 109th Congress*, by Keith Bea.

[64] The Hazardous Materials Emergency Preparedness Fund is authorized at 49 U.S.C. 5116(i).

[65] U.S. Department of Education, Office of Safe and Drug-Free Schools, "Overview Information, Emergency Response and Crisis Management Grant Program; Notice Inviting Applications for New Awards for Fiscal Year (FY) 2005," *Federal Register*, vol. 70, June 21, 2005, pp. 35649-35655.

[66] This report does not examine response teams primarily composed of non-federal personnel but supported with federal funds, notably the Metropolitan Medical Response System (MMRS) teams, discussed at [https://www.mmrs.fema.gov/default.aspx], and Urban Search and Rescue teams, discussed at [http://www.fema.gov/usr/], both visited Aug. 18, 2005.

[67] U.S. General Accounting Office, *Combating Terrorism: Federal Response Teams Provide Varied Capabilities; Opportunities Remain to Improve Coordination*, GAO Report GAO-01-14 (Washington: Nov. 30, 2000), p. 4. The data in this report have not been updated by GAO.

[68] Table 3 constitutes a partial listing of federal response teams and coordination centers. Some are not listed because they are region specific and operated by one agency, such as the National Capital Response Team operated by the FBI and the teams established by DOD for the National Capital Region through Joint Task Forces East, West, or Headquarters-National Capital Region, in the Washington, DC area. Others, such as the Coast Guard Strike Force, the Army's Technical Escort Unit, and the Air Force Disaster Preparedness/Full Spectrum Threat Response community are not listed because scant information on those entities is generally available.

[69] The Federal Interagency Committee on Emergency Medical Services (FICEMS), chaired by U.S. Fire Administration officials, "serves as a forum to establish and facilitate effective communications and coordination between and among Federal departments and agencies involved in activities related to EMS." See FICEMS website at [http://www.usfa.fema.gov/subjects/ems/ficems.shtm], visited Aug. 4, 2005. For background on NDMS, see CRS Report RL31719, *An Overview of the U.S. Public Health System in the Context of Emergency Preparedness*, by Sarah A. Lister.

[70] 42 U.S.C. 247d.

[71] See "Emergency Preparedness and Response" at [http://www.bt.cdc.gov/], visited June 30, 2005.

[72] U.S. Department of Health and Human Services, Office of the Secretary, "Office of Public Health Emergency Preparedness Statement of Organization, Functions, and Delegations of Authority," *Federal Register*, vol. 70, Feb. 1, 2005, pp. 5183-5184. For information on NDMS, see "National Disaster Medical System" at the HHS website, [http://www.ndms.dhhs.gov/], visited Feb. 1, 2005. The seal of DHS appears on the NDMS website. MMRS capabilities are discussed at [https://www.mmrs.fema.gov/default.aspx], visited Aug. 12, 2005.

[73] General information about these components is presented in CRS Report RL30802, *Reserve Component Personnel Issues: Questions and Answers*, by Lawrence Kapp. For an overview of basic duties of the components, see remarks of Peter Verga, Deputy Assistant Secretary of Defense before the Defense Forum Foundation Forum on Homeland Defense, available at *CQ.com*, June 17, 2005; available by subscription.

[74] Title 10 authority related to emergency response needs in the United States includes the provision of military aid to state governments in order to suppress insurrections (10 U.S.C. 331, 333) and assisting civilian agencies (10 U.S.C. 372-374). Regulations (32 CFR Part 501) specify the emergency conditions under which aid may be provided to civil authorities. Such aid, however, cannot compromise military preparedness (10 U.S.C. 376). Reserve forces may be called to active duty by the President "at time of national emergency" (10 U.S.C. 12301-12302), but cannot be called to assist "in time of a serious natural or manmade disaster, accident, or catastrophe" or insurrection (10 U.S.C. 12304(b)).

[75] The "Militia of the several states" may be "called into the actual Service of the United States" by the President (U.S. Constitution, Art. II, Sec. 2). Congress is authorized "To provide for calling forth the Militia to execute the Laws of the Union, suppress Insurrections and repel Invasions" (U.S. Constitution, Art. I, Sec. 8). Federal law also provides for the federalization of the Guard [10 U.S.C. 12406]. National Guard members called to active duty status are relieved from their state (or insular area) military role unless the President authorizes service in both duty statuses and the governor consents to such dual service (32 U.S.C. 325). President Bush delegated to the Secretary of Defense the authority to approve such dual status for the National Guard involved in security for the Group of Eight (G8) Summit in June, 2004. See U.S. President, "Command and Control of National Guard for 2004 Group of Eight ("G8") Summit, *Federal Register*, vol. 69, June 10, 2004, p. 32831-32833.

[76] See National Academy of Public Administration, *The Role of the National Guard in Emergency Preparedness and Response* (Washington: 1997).

[77] The term is defined as "an activity undertaken for the military protection of the territory or domestic population of the United States, or of infrastructure or other assets of the United States determined by the Secretary of Defense as being critical to national security, from a threat or aggression against the United States." 32 U.S.C. 901(1).

[78] U.S. Department of Defense, "Military Support to Civil Authorities (MSCA)," DOD Directive Number 3025.1, Jan. 15, 1993; "Military Assistance to Civil Authorities," DOD Directive Number 3025.15, Feb. 18, 1997.

[79] See testimony of General Richard A. Cody, U.S. Army, before U.S. Congress, House Armed Services Committee, Subcommittee on Military Personnel, *Army Force Strength*, hearing, 109th Cong., 1st sess., Feb. 2, 2005 (not yet printed).

[80] Ibid.

[81] Commonly referred to as the Posse Comitatus Act, the provision prohibits military activity that could intrude on civil authorities unless specifically authorized. 18 U.S.C. 1385. See CRS Report RS21012, *Terrorism: Some Legal Restrictions on Military Assistance to Domestic Authorities Following a Terrorist Attack*, by Charles Doyle and Jennifer Elsea.

[82] The perspectives of local government officials regarding the role of the Department of Defense are presented in International Association of Emergency Managers, "Special Focus Issue: Defense Support to Civil Authorities," *IAEM Bulletin*, vol. 22, July 2005, entire issue.

[83] U.S. Department of Defense, *Joint Publication 3-26 Homeland Security* (Washington: 2005), pp. IV-4 through IV-7, available at [http://www.fas.org/irp/doddir/dod/jp3_26.pdf], visited Aug. 18, 2005. The doctrine notes that responsibility for assistance for civil disturbances rests with the Attorney General.

[84] Issues associated with the homeland security response roles of the Coast Guard, part of DHS, and the U.S Navy, part of DOD, are discussed in CRS Report RS21230, *Homeland Security: Navy Operations — Background and Issues for Congress*, by Ronald O'Rourke.

[85] For background and discussion of the NORTHCOM role, see CRS Report RL31615, *Homeland Security: The Department of Defense's Role*, by Steve Bowman. The civil support role of the Department of Defense is discussed in U.S. Department of Defense, *Strategy for Homeland Defense and Civil Support* (Washington: 2005), available at [http://www.defenselink.mil/news/Jun2005/d20050630homeland.pdf], visited July 1, 2005.

[86] Statement of James Carafano, *Outlook for the Department of Homeland Security*, Jan. 26, 2005.

[87] Information about NORTHCOM is available at [http://www.northcom.mil/index.cfm?fuseaction=news.factsheets#usnorthcom], visited Aug. 18, 2005.

[88] Bradley Graham, "War Plans Drafted to Counter Terror Attacks in U.S.," *The Washington Post*, Aug. 8, 2005, pp. A1, A7.

[89] U.S. Department of Defense, *Joint Publication 3-26 Homeland Security* (Washington: 2005), available at [http://www.fas.org/irp/doddir/dod/jp3_26.pdf], visited Aug. 18, 2005.

[90] Ibid., p. ii.

[91] Department of Homeland Security, Office of Inspector General, *Major Management Challenges Facing the Department of Homeland Security*, OIG-05-06 (Washington: 2004), p. 1.

[92] Secretary Chertoff letter, p. 4.

[93] Statement of James Carafano, The Heritage Foundation, in *Outlook for the Department of Homeland Security*, Jan. 26, 2005. For information on the proposal, see Alan Kochems, "Who's on First? A Strategy for Protecting Critical Infrastructure," *Backgrounder*, No. 1851 (Washington: The Heritage Foundation, 2005).

[94] Ibid., statement of Richard Skinner, Department of Homeland Security.

[95] Robert Block, "Homeland Security Wrestles with Revamp," *The Wall Street Journal*, June 13, 2005, p. A4.

[96] David E. Liebersbach, NEMA President, to Honorable Susan Collins and Honorable Joe Lieberman, July 27, 2005.

[97] Amy E. Smithson and Leslie-Anne Levy, *Ataxia: The Chemical and Biological Terrorism Threat and the U.S. Response* (Washington: The Henry L. Stimson Center, Oct. 2000), p. xv.

[98] 42 U.S.C. 5143.

[99] U.S. Department of Homeland Security, *National Response Plan* (Washington: 2004), p. 33.

[100] During the Clinton Administration, the Director of FEMA was accorded cabinet level status through an executive decision made by President Clinton. The recommendation to return responsibility for emergency management to the top levels of government was included in National Academy of Public Administration, *Coping with Catastrophe: Building an Emergency Management System to Meet People's Needs in Natural and Manmade Disasters* (Washington, 1993).

[101] Some congressional and administrative officials contended after the lessons learned from Hurricane Andrew in 1992 that a federal 911 emergency response capability had to be built. The establishment of the 24/7 Homeland Security Operations Center within DHS constitutes the observation and "dispatch" functions associated with a national 911 capability. In contrast is the contention in President Carter's 1978 reorganization plan that "there is no need to develop a separate set of federal skills and capabilities *for those rare occasions when catastrophe occurs*" [emphasis added]. U.S. President (Carter), "Reorganization Plan No. 3 of 1978," H. Doc. 95-356, 95th Cong., 1st sess. (Washington: GPO, 1978), p. 3.

[102] Lee Clarke, "Worst-Case Thinking: An Idea Whose Time Has Come," *Natural Hazards Observer*, vol. XXIX, Jan. 2005. Based on a forthcoming text, *Worst Cases*, scheduled to be published by the University of Chicago Press.

[103] Remarks of Secretary Chertoff in U.S. Congress, House Homeland Security Committee, *Review of Department of Homeland Security Organization* (Washington: 2005), transcript available by subscription through CQ Homeland Security.

[104] Ibid.

[105] U.S. Congress, House Committee on Transportation and Infrastructure, Subcommittee on Economic Development, Public Buildings and Emergency Management, *The Fiscal Year 2005 Budget Request for the Department of Homeland Security's Emergency Preparedness and Response Directorate, the Office for Domestic Preparedness, and First Responder Funding*, hearing, 108th Cong., 2nd sess., Mar. 18, 2004 (Washington: GPO, 2005), pp. 4-7.

[106] U.S. Government Accountability Office, *Homeland Security: DHS' Efforts to Enhance First Responders' All-Hazards Capabilities Continue to Evolve*, GAO Report GAO-05-652, (Washington: July 11, 2005).

[107] Hearings were held in the Senate and the House on the proposal shortly after it was released. Also, a report issued by the Democratic members of the House Homeland Security Committee shortly after the Secretary released the findings raises questions about the effect of the proposal on different preparedness needs and constituencies, the absence of information on communications interoperability, and training. See U.S. House of Representatives, *Protecting America Against Terrorists: The Case for a Comprehensive Reorganization of the Department of Homeland Security* (Washington: 2005).

[108] Maintaining effective relations with non-federal partners involved in the delivery of homeland security and emergency management services may be one of the paramount elements of the debate. As noted in Appendix B of this report, dissatisfaction of state officials with the reorganization plan developed by President Nixon in 1973 led to calls

for a new approach that culminated in the establishment of FEMA in 1978. See page 59 of this report.

109 Legislation before the 109th Congress (H.R. 3477) would direct the DHS Secretary to report to Congress on a plan to establish consolidated and co-located regional offices throughout the nation.

110 Letter of Secretary Chertoff to Congress, July 13, 2005, p. 5.

111 U.S. Government Accountability Office, *Continuity of Operations: Improved Planning Needed to Ensure Delivery of Essential Government Services,* GAO-04-160, Feb. 27, 2004. See also U.S. Government Accountability Office, *Continuity of Operations: Agency Plans Have Improved, but Better Oversight Could Assist Agencies in Preparing for Emergencies,* GAO-05-577, Apr. 28, 2005. U.S. Congress, House Committee on Government Reform, *Can Federal Agencies Function in the Wake of a Disaster? A Status Report on Federal Agencies' Continuity of Operations Plans*, hearing, 108th Cong., 2nd sess., Apr. 22, 2004 (Washington: GPO, 2004).

112 For background, see U.S. Nuclear Regulatory Commission, "Announcement of a Public Meeting to Discuss Selected Topics for the Review of Emergency Preparedness (EP) Regulations and Guidance for Commercial Nuclear Power Plants," *Federal Register*, vol. 70, July 28, 2005, pp. 43721-43725.

113 Legislation (H.R. 3477) to establish a deadline for submission of a plan for establishing and co-locating regional offices is pending before the 109th Congress.

114 49 U.S.C. 5101 et seq.

115 One might argue, however, that the use of FEMA resources in 2003 to locate the pieces of the space shuttle *Columbia* indicated a need for broader response capabilities.

116 U.S. President (Eisenhower), "Reorganization Plan 1 of 1958," *Federal Register*, vol. 23, July 1, 1958, p. 4991.

117 National Commission on the Public Service, *Urgent Business for America: Revitalizing the Federal Government for the 21st Century* (Washington: The Brookings Institution, 2003), available at [http://www.brookings.edu/gs/cps/volcker/urgentbusinessreport.htm], visited June 16, 2005.

118 See CRS Report RL32851, *Hazardous Material Transportation Security: Highway and Rail Modes*, by Paul F. Rothberg. The surface transportation legislation passed by Congress and approved by the President provides new authority for the Secretary of Transportation regarding the transportation of hazardous material, including the establishment of a Hazardous Material Emergency Preparedness Fund for grants to state and local first responders. See 49 U.S.C. 5102 et seq.

119 For information on the recommendation see *Back to the Future: An Agenda for Federal Leadership of Emergency Medical Services* , available at [http://www.homelandsecurity.gwu.edu/reports/HSPI_EMS_task_force_report_5-2-05.pdf], visited June 13, 2005.

120 See Eileen Sullivan, "Tension Between Emergency Medical Crews and Firefighters Stays on Simmer," *CQ Homeland Security*, June 10, 2005, see Eileen Sullivan, "Study: Emergency Medical Services Should Move from DOT to DHS," *CQ Homeland Security*, May 2, 2005

121 "Secretary Michael Chertoff U.S. Department of Homeland Security Second Stage Review Remarks," available at [http://www.dhs.gov/dhspublic/display?content=4597], visited Aug. 11, 2005.

[122] Remarks of Secretary Chertoff, U.S. Congress, House Homeland Security Committee, *Review of Department of Homeland Security Organization* (Washington: 2005), transcript available by subscription through CQ Homeland Security.

[123] For example, see Jeff Nesmith, "Who's In Charge if Bird Flu Strikes — Docs or Cops?," *The Atlanta Journal-Constitution*, Aug. 24, 2005, accessed at [http://www.ajc.com/today/content/epaper/editions/today/news_34c09175043fd0ef10d0.html], visited Aug. 30, 2005.

[124] Ari Schuler and others, "Executive Government Positions of Influence in Biodefense: The Bio-Plum Book," *Biosecurity and Bioterrorism: Biodefense Strategy, Practice, and Science*, vol. 2, no. 4, 2004, available at [http://www.upmc-biosecurity.org/pages/publications/articles.html#2004], visited June 10, 2005.

[125] Statement of James Carafano, *Outlook for the Department of Homeland Security*, Jan. 26, 2005.

[126] Statement of Dr. Julie L. Gerberding before the House Homeland Security Committee, Subcommittee on Prevention of Nuclear and Biological Attack, July 28, 2005, transcript available through [CQ.com subscription]. At the same hearing, Dr. John Vitko, Jr., Director of the Biological Countermeasures Portfolio in DHS, identified six units within the department and eight entities outside DHS that "have major roles and responsibilities in implementing the national biodefense strategy."

[127] U.S. Congress, House Government Reform Committee, Subcommittee on National Security, Emerging Threats and International Relations, "Elusive Antidotes: Progress Developing Chemical, Biological, Radiological and Nuclear Countermeasures," June 14, 2005, not yet printed.

[128] "The primary mission of the Department is to — (D) carry out all functions of entities transferred to the Department, including by acting as a focal point regarding natural and manmade crises and emergency planning..." Sec. 101(a) of P.L. 107-296, 6 U.S.C. 111(b)(1).

[129] For background, see CRS Report RL32803, *The National Preparedness System: Issues in the 109th Congress*, by Keith Bea.

[130] GAO, *Continuity of Operations: Improved Planning Needed to Ensure Delivery of Essential Government Services*, p. 9.

[131] Darren Goode, "Lawmaker Pushes Use of Military Technology in Disaster Response," *Congress Daily*, Jan. 10, 2005.

[132] For information on NORTHCOM, see CRS Report RS21322, *Homeland Security: Establishment and Implementation of Northern Command*, by Christopher Bolkcum and Steve Bowman.

[133] R. Steven Daniels and Carolyn L. Clark-Daniels, Transforming Government: The Renewal and Revitalization of the Federal Emergency Management Agency, (Birmingham, AL: Department of Government and Public Service, University of Alabama, 2000), pp. 7-8. This finding resembles one reached by other researchers who examined six public sector mergers. One of the studies led to the conclusion that a clear concept of mission and desired results, and a merger of common missions, should be "articulated from the start and [drive] the merger's implementation." See Peter Frumkin, *Making Public Sector Mergers Work: Lessons Learned* (Cambridge, MA: John F. Kennedy School of Government, Harvard University, 2003), p. 22.

[134] Ibid, p. 7.

[135] President George W. Bush, *Homeland Security Presidential Directive-1*, (Washington: 2001), available at [http://www.whitehouse.gov/news/releases/2001/10/20011030-1.html], visited June 16, 2005.

[136] See CRS Report RL31357, *Federal Interagency Coordinative Mechanisms: Varied Types and Numerous Devices*, by Frederick M. Kaiser.

[137] P.L. 99-433, 100 Stat. 992-1075b.

[138] Gordon Nathaniel Lederman, *Reorganizing the Joint Chiefs of Staff: The Goldwater-Nichols Act of 1986* (Westport, CT: Greenwood Press, 1999).

[139] Under current law, the President must appoint a federal coordinating office (FCO) after a major disaster declaration is issued. The FCO must ensure that the assistance provided by federal and non-federal agencies to the disaster-stricken area is delivered in a coordinated fashion. See 42 U.S.C. 5143. See also the requirements for combatant commanders at 10 U.S.C. 164.

[140] Section 102 of the act sets forth the functions of the Secretary. The statute authorizes the Secretary to coordinate certain activities — preparedness with non-federal entities, federal communications related to homeland security, and the distribution of warnings. See 6 U.S.C. 112.

[141] Peter Szanton, *Federal Reorganization: What Have We Learned* (Chatham, NJ: Chatham House Publishers, 1981).

[142] Michele L. Landis, "Let Me Next Time Be Tried by Fire: Disaster Relief and the Origins of the American Welfare State 1789-1874," *Northwestern University Law Review*, vol. 92, spring 1998, pp. 967-1034. A list of disaster legislation enacted by Congress from 1803 through 1943 may be found in Rep. Harold Hagen, Statement for the Record, *Congressional Record*, vol. 96, Aug. 7, 1950, pp. 11900-11902.

[143] The exception to this general statement concerns flood prevention policies enacted since the late 19th century. See CRS Report RL32972, *Federal Flood Insurance: The Repetitive Loss Problem*, by Rawle King.

[144] "To authorize federal assistance to states and local governments in major disasters, and for other purposes," P.L. 81-875, 64 Stat. 1109-1111. This legislation is the forerunner of the considerably wider authority granted the President in the Robert T. Stafford Disaster Relief and Emergency Assistance Act, 42 U.S.C. 5121 et seq.

[145] For a chronology of the administrative assignments, see CRS Report RL31510, *Proposed Transfer of FEMA to the Department of Homeland Security*, by Keith Bea (available from the author).

[146] In addition to enlarging the types of disaster relief to be provided, legislation enacted during these years established new federal policy to help state and local governments prepare for emergencies, recover from disasters, and reduce losses (mitigate hazards) anticipated from future disasters. Legislation was enacted in 1966 (P.L. 89-796, 80 Stat. 1316 et seq.), 1969 (P.L. 91-79, 83 Stat. 125 et seq.), 1970 (P.L. 91-606, 84 Stat. 1744 et seq.), and 1974 (P.L. 93-288, 88 Stat. 143 et seq.).

[147] U.S. Office of Management and Budget, President's Reorganization Project, *Federal Emergency Preparedness and Response Historical Survey* (Washington: 1978), p. 40.

[148] The CEM concept largely derived from conversations held among state officials in the 1970s. See National Governors' Association, *Comprehensive Emergency Management: A Governor's Guide*, (Washington: 1979). The Council of State Governments,

Comprehensive Emergency Preparedness Planning in State Government (Lexington, KY: 1976).

149 U.S. President (Carter), "Reorganization Plan No. 3 of 1978," H. Doc. 95-356, 95th Cong., 1st sess. (Washington: GPO, 1978), pp. 2-3.

150 In 1976 Congress amended the Civil Defense Act of 1950 to recognize that the civil defense system could be used "to provide relief and assistance to people in areas of the United States struck by disasters other than disasters caused by enemy attack." P.L. 94-361, 90 Stat. 931.

151 For details, see U.S. Federal Emergency Management Agency, *Guide for All-Hazard Emergency Operations Planning: State and Local guide (101)* (Washington: 2001), available at [http://www.fema.gov/rrr/allhzpln.shtm], visited Aug. 19, 2005.

152 U.S. President Carter, "Reorganization Plan No. 3 of 1978," p. 3.

153 U.S. Office of the President, Office of Management and Budget, President's Reorganization Project, *Federal Emergency Preparedness and Response Historical Survey* (Washington: 1978), pp. 69-70.

154 A few of the many reports issued by the General Accounting Office (now Government Accountability Office) in the 1980s and 1990s provide an indication of the challenges faced by FEMA administrators. See U.S. General Accounting Office, *Federal Disaster Assistance: What Should the Policy Be?*, GAO Report PAD-80-39 (Washington: June 16, 1980). *Stronger Direction Needed for the National Earthquake Program*, GAO report RCED-83-103 (Washington: July 26, 1983). *Management of the Federal Emergency Management Agency — A System Being Developed*, GAO Report GGD-83-9 (Washington: January 6, 1983). *Disaster Assistance: Federal, State, and Local Response to Natural Disasters Needs Improvement*, GAO Report RCED-91-43 (Washington: March 6, 1991). *Disaster Management: Improving the Nation's Response to Catastrophic Disasters*, GAO Report RCED-93-186 (Washington: July 23, 1993).

155 See S. 1697, 103rd Cong., the Federal Disaster Preparedness and Response Act of 1994.

156 For a discussion of administration and policy changes made see R. Steven Daniels and Carolyn L. Clark-Daniels, *Transforming Government: The Renewal and Revitalization of the Federal Emergency Management Agency* (The PricewaterhouseCoopers Endowment for the Business of Government: University of Alabama at Birmingham, 2000). FEMA exercised limited authority regarding human caused catastrophes such as terrorist attacks. Presidential Decision Directive-39 authorized FEMA to undertake consequence management tasks associated with the use of weapons of mass destruction, and the agency had a limited role in training state and local officials to respond to the use of such weapons. For an overview of FEMA's authority under PDD-39, see the 1997 hearing statement of Catherine H. Light, Director, FEMA Terrorism Coordination Unit, before the House National Security Committee, available at [http://www.fas.org/spp/starwars/congress/ 1997_h/h971104l.htm], visited Aug. 16, 2005.

157 For example, a review of federal emergency authorities related to the release of hazardous material found the existing system to be "complex, confusing, and costly." This conclusion led to the development of an Integrated Contingency Plan Guidance "to be used by facilities to prepare emergency response plans." U.S. Environmental Protection Agency, "The National Response Team's Integrated Contingency Plan Guidance," *Federal Register*, vol. 61, June 5, 1996, pp. 28642-28664.

[158] U.S. Commission on National Security/21st Century, *Road Map for National Security: Imperative for Change*, available at [http://www.au.af.mil/au/awc/awcgate/nssg/phase IIIfr .pdf], visited July 28, 2005.

[159] Smithson and Levy, *Ataxia: The Chemical and Biological Terrorism Threat and the U.S. Response*, p. xiv.

[160] The 108th Congress reemphasized the need to ensure coordination through enactment of Sec. 7405 of the Intelligence Reform and Terrorism Prevention Act of 2004 (P.L. 108-458) as follows: "The Secretary of Homeland Security shall ensure that there is effective and ongoing coordination of federal efforts to prevent, prepare for, and respond to acts of terrorism and other major disasters and emergencies among the divisions of the Department of Homeland Security, including the Directorate of Emergency Preparedness and Response and the Office for State and Local Government Coordination and Preparedness."

INDEX

I

J

L

M

N

O

P

U

V

W